Once Upon an Island

Once Upon an Island

BY DAVID CONOVER

To my teacher Allan Dartford with gratitude and affection

I thank Anne C. Ambrose for her encouragement and criticism

SAN JUAN PUBLISHING
P.O. Box 923
Woodinville, WA 98072
425-485-2813
sanjuanbooks@yahoo.com

Publisher: Michael D. McCloskey
Designer and Mapmaker: Jennifer LaRock Shontz
Proofreader: Sherrill Carlson
All photographs taken by the author unless otherwise noted.

Front cover photograph: *David and Jeanne unloading Bertha at the campsite where they settled for the first few days.*
Back cover photograph: *Mr. & Mrs. Crusoe*
Frontispiece: *Looking north toward Wallace Island with Conover Cove located center left.*

San Juan Publishing would like to thank Barbara Conover for granting permission to reprint this edition of *Once Upon an Island.*

ISBN 0-9707399-1-5

Library of Congress Catalog Card Number: 2003097220

First Printing November 2003
10 9 8 7 6 5 4 3 2 1
Printed in Canada

Contents

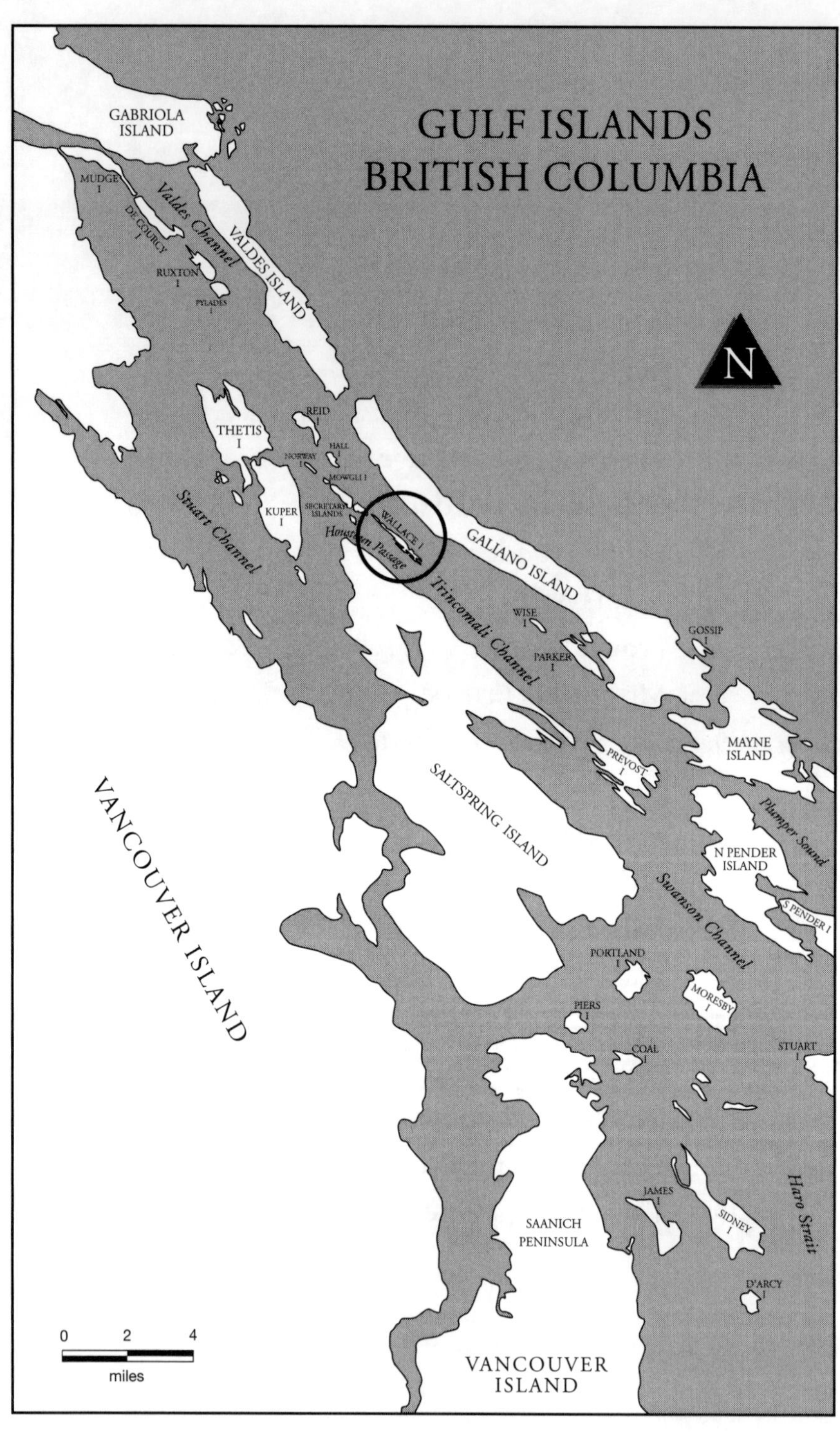
GULF ISLANDS
BRITISH COLUMBIA
N
GABRIOLA ISLAND
MUDGE I
DE COURCY I
Valdes Channel
VALDES ISLAND
RUXTON I
PYLADES I
THETIS I
REID I
HALL I
NORWAY I
MOWGLI I
KUPER I
SECRETARY ISLANDS
WALLACE I
Houstoun Passage
Stuart Channel
GALIANO ISLAND
Trincomali Channel
WISE I
PARKER I
GOSSIP I
PREVOST I
MAYNE ISLAND
SALTSPRING ISLAND
Plumper Sound
N PENDER ISLAND
S PENDER I
Swanson Channel
VANCOUVER ISLAND
PORTLAND I
MORESBY I
PIERS I
COAL I
STUART I
JAMES I
SIDNEY I
Haro Strait
SAANICH PENINSULA
D'ARCY I
0 2 4
miles
VANCOUVER ISLAND

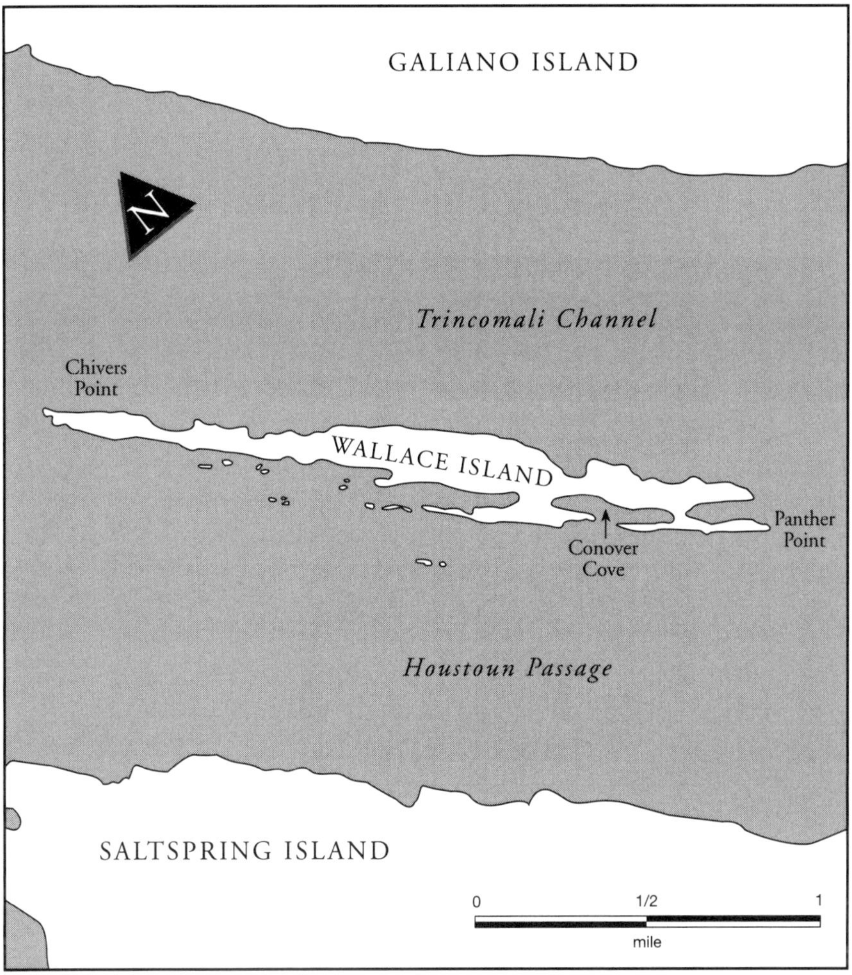

GALIANO ISLAND

N

Trincomali Channel

Chivers Point

WALLACE ISLAND

Panther Point

Conover Cove

Houstoun Passage

SALTSPRING ISLAND

0 1/2 1
mile

The events and settings of this story are real. Except where friends have given me permission to use their names, all other characters are fictitious.

To live outdoors with the woman a man loves is of all lives the most complete and free.

—R. L. Stevenson

CHAPTER 1

Island Fever

IF, BY CHANCE, YOU HAVE LIVED AN ADVENTUROUS LIFE on a tiny uninhabited fir-clad island along the wind-swept shores of British Columbia for almost twenty years, you are bound to wake up some morning and wonder how it all began.

The other morning, I awoke early and began to think about it. How wonderful it is to realize a dream while you are young, particularly when your dream is to live on your own island. And how intensely satisfying this life is. You can tell the world to go to blazes, or take from it what you wish. What a strange sort of existence!

I rolled over in bed. "Jeanne, how the devil did we get here?"

She thought for a moment. "It was your dream, remember? I just came along for the ride. But I love it."

She was right. That was the way it started—the day I noticed an ad in the college newspaper: "Summer counselor wanted. Boys' camp, 6 weeks, expenses paid, on a Canadian Island." My curiosity took me to Canada that summer. Then it began—my love affair with an island.

Or perhaps it really began that crystal April morning in 1946 when we shoved off in a borrowed rowboat from an abandoned wharf at Fernwood, British Columbia, into a vast inland sea of islands to explore one in particular.

"You sure we're rowing toward the right island?" Jeanne shifted her gaze toward the bow, suspiciously eyeing the green wall of firs that loomed up ahead.

"I haven't forgotten. That's Wallace all right."

"But we're almost there. Where's the hidden cove you told me about?" She was especially pretty when she was disturbed.

"It's a real corker to find," I admitted, slowing on the oars. "The entrance is so well camouflaged, few people know it's there."

I knew we would soon be approaching the giant kelp beds, a ring of razor-sharp reefs that guard the entrance of the cove. The oars creaked as the boat leaped forward, skimming the sea toward the tree-studded shoreline. Buffleheads skittered the surface ahead of us, crash-diving and showing us their fluffy white tails. For a long time, we seemed to get nowhere. But gradually the old wharf on Salt Spring Island began to fade behind us. My shoulders were aching, and I stretched my toes to ease my cramped legs. I had dreamed of this moment so often—of showing the island to Jeanne—that now, after ten years, the excitement of seeing it again canceled out all my discomfort.

The boat entered the dense kelp. The oars came up heavy, tangled with its slippery brown arms, while under us the dark ruffled fingers pointed to menacing rocks below.

Jeanne peered over the side, fascinated. This was the first time in all her twenty-four years she had ever been on the sea. "I hope you know where you're going," she worried.

I nodded reassuringly. Moments later the reef lay behind us, and I found myself straining on the oars. My watch showed ten o'clock. It was unbelievable that two miles would take this long to row. Distances among these islands were deceiving. And I began to wonder whether the sea would remain calm for us to return by nightfall.

In the morning there were ferries to catch, a 1,200-mile drive home to Los Angeles, and a job at North American Aviation waiting for me. I hated the thought of going back to the grind. Even the war, still fresh in my mind, had not erased the memory of the cold stare of a timeclock or the soul-destroying roar of an aircraft plant. And as I heaved on the oars, I wondered if there wasn't more in a man than what he knows how to do. Was I destined to fall back and become a cog in a machine once more? Would I, as before, find all my working hours fringed with boredom, empty of meaning, forever waiting for the quitting whistle, a raise, holidays, with nothing in life to live for except a paycheck? What was it inside me, I wondered, that demanded a challenge from life?

Two hundred yards from shore, green slopes suddenly mirrored on the sea. A pleased smile crept across Jeanne's oval face. She laid her hands on mine and we rowed in unison, stroking and looking, without saying a word. The island crept closer, filling our lungs with the sweet scent of evergreens as we headed toward the slight depression among the highest firs. Below them, a sandstone ridge honeycombed with little caves ran the length of the island like a rampart defending its shores. Ahead, now, the ridge gradually dipped as though it slumped from the weight of red-faced madronas, twisted oaks, and tall spiraling firs.

Jeanne pointed. "Look, there's the opening."

A gap yawned in the sandstone. Then the island closed around us, and we were engulfed by the forest in a motionless landlocked lagoon a half-mile long and several hundred yards wide, an exquisite miniature of San Francisco Bay.

The cove nestled between gentle wooded slopes, guarded by a wall of firs and cedars that grew to enormous heights. Madronas, maples, and alders dotted the banks like colored crayons, their reflection creating the incredible illusion that our boat was stranded in the trees. Between outcrops of sandstone, white shell beaches glistened in the sun. From the woods came the fluttering of wings and the cries of strange birds. Eagles wheeled overhead. On one leg, a great blue heron watched us with a cynical eye. A mink hobbled along the rocks. With gutteral noises, a mother prodded her ducklings ashore into the safety of the forest. Schools of needlefish churned arcs of silver. Ducks and cormorants floated everywhere. It was as though we had stumbled onto a lost world.

"Gosh, Dave"—Jeanne's voice trembled—"how could anything be so beautiful?"

"It's even lovelier than I remembered."

We glided deep into the cove, devouring this unreal world as if we were visitors from another planet, not letting a scent, a sight, or sound escape us, flinging our hearts open to the island's charm. The air held the exotic tang of seaweed and cedar drying in the sun. We breathed deeply of the strange aroma as our eyes probed the lush greenery and almost tropic-like growth. What perfect quiet!

Now and then gazing at the colorful marine garden blooming beneath us, we edged toward the nearest beach, slicing through patches of sea lettuce and startling flounder and cod from their naps. Curled madrona leaves whisked by us like proud Spanish galleons.

We began to lose all sense of time. The sun climbed higher, spinning ribbons of light on maidenhair fern, thickets of salal, and Oregon grape fighting for toeholds along the shadowy bank. Dogwood blossomed in dark glens and pink lady-slippers carpeted the feet of baby firs. From pockets of earth in the bedrock sprang the orange flame of Indian paintbrush, wild peas, and lupine. The rarest perfume could not match the fragrance in the air.

How different this was from the barren brown hills of Los Angeles. April had always been just another month. Now, for the first time, we thrilled to spring as though we had discovered something precious and rare. And we began to feel a little sad, wondering how much we had missed. Just then, our cockleshell glided onto a powdery beach.

"This is it, Jeanne," I announced. "Wallace Island."

We pulled the boat up onto the beach and stood silent.

"I never knew anything like this existed except in storybooks," Jeanne said finally. "Now I know what prompted your beautiful love letters years ago."

We spent the day roaming the island, exploring every wooded trail and inlet, scrambling across deep valleys of salal and poking our noses in every cave and crevice. We fought and clawed our way up sandstone ridges, sliding down slopes on sleds of moss and madrona leaves. Hand in hand, we roamed among stately groves of alders and maples, more than convinced that this was paradise.

When we came to the small clearing near the cove, it looked vaguely familiar to me.

"Dave, come here." Jeanne motioned to something on the ground. "What's this?"

Brushing aside the tall grass, I peered down at a raft of decayed boards. "Our old tent platform—where we camped the summer of '36."

It was then I had first visited Wallace, and here I had written my first love letters. Beside me stood the girl I'd written them to, now my wife. I glanced at her fondly. What a cinch it would be to write them again.

Our love affair blossomed when we were fifteen. Jeanne was the prettiest girl in Los Angeles High. She walked down the school corridors like a queen with a train of knights in tow. Few boys could match her enthusiasm, her zest for life, and her innate talent of doing everything she did well. In college we eloped to Wickenburg, Arizona, with neither time nor money for a honeymoon. Instead, I plunged into photography, working days at North American and studying commercial photography at night school. The few moments we shared, I shot pictures of Jeanne. With a cover girl like her, how could a photographer fail?

Then came Pearl Harbor. Guard duty, scrubbing latrines, combat photography, the South Pacific. When it was all over, we decided we owed each other a honeymoon trip. "Let's visit Wallace Island," Jeanne suggested. "It's all I've heard you talk about for years."

Only this morning, when we borrowed the rowboat at Fernwood from Vic Bettiss, the caretaker of the old boys' camp, did we learn that the camp had long since been abandoned.

The camp buildings had been attacked by an expert scavenger. Floors, ceilings, even the walls, were stripped bare.

"This was the messhall"—I nodded at the larger structure, which looked like an open-air pavilion—"and that's the cookhouse." We marveled how the roofs had lasted, perched on log posts spaced so far apart.

Through waist-high salal, we returned to the clearing and looked back at the buildings. Like ancient pagodas swamped by the jungle, they gave an air of mystery to the old campsite, and as we turned to go, a cool breeze rustled across the tall grass and brushed my cheek like a human hand. For a second, I thought I could faintly hear the shouts of rollicking boys, fading into squeals of laughter.

Shadows began to skirt the lonely clearing, and we knew our precious time was running short. What a shame to worry about the clock, here on an island where time mattered so little. How wonderful to have a world like this for one's own!

Under the shade of a giant madrona, we ate our peanut butter and jelly sandwiches, washing them down with a thermos of Vic's fresh milk. Any similarity to city milk, we agreed, was purely accidental.

"Where did Vic stay?" Jeanne asked.

I had forgotten about the caretaker's dwelling. Vic had patched up an old cabin built long ago by the island's first owner. While we searched for it, prowling through the bush and brown grass, I marveled at how the island had changed. The clearing I remembered had been three times as large and dotted with dozens of apple trees, cutting a green swath from the cove to a bay on the opposite side. An invasion of firs had shrunk it to a couple of acres, and only a handful of bewildered apple trees remained. The pink scented buds that loaded their brave shaggy limbs told us they were still alive.

"All this was open, in short grass, with nothing but apple trees," I explained, pointing in a broad semicircle. I couldn't believe nature had been so prolific.

"You forget, darling—this isn't our backyard cubicle on West Vernon. For us Californians to make anything grow, it has to be watered, fertilized, sprayed—"

"Here, I suppose, you drop in a seed and it pops up and tickles your nose next morning."

Jeanne broke into a smile. She understood what I did not have the courage to admit: I was now as much of a stranger to the island as she.

We finally discovered the cabin near the cove.

"Why, it's just an old shack," Jeanne said. She looked out over the water and shook her head. "To think this lovely view goes to waste while we stare into old lady Crawford's kitchen every morning."

The shack reminded us of a hobo shanty. Its sides were planked with driftwood boards, and rags covered the two tiny broken windows. It looked sad and forlorn, as if the only thing holding it upright was the spidery arms of the apple tree hugging the mossy roof.

We pushed open the handmade door. The floor creaked and bent under us, and foul air filled our lungs. The bleak twelve-by-fourteen room made us shudder. We half-expected some bearded old codger to stalk out of the shadows. Cobwebs, rags, and bits of fish netting hung from the open rafters. A rusty wood stove peered at us under a window, and an aged brown sink jutted out from the bare shiplapped wall. Jeanne shrank back from the sink when she noticed it was full of spiders curled up into little black balls. Beside a rough wooden table, a bed made from driftwood and gunny sacks lay collapsed on the floor amid fragments of rose-colored dishes.

We tiptoed back toward the door, catching the sound of a rat gnawing inside the wall. On a dusty shelf loaded with bottles, pots, rusty forks, and tobacco tins was a leatherbound Bible. Blowing the dust from it, I opened it carefully while Jeanne looked over my shoulder. On the flyleaf scrawled in firm pencil were the words: "Jack Chivers, 1898." The name struck a familiar note. Curiously I thumbed through the pages for other signs of writing or documents.

"Wait a minute," Jeanne cried. "Turn back a few pages. That's it—there."

Underscored in pencil was a passage from Matthew. "For where your treasure is, there will your heart be also."

Suddenly I began to remember the stories Vic had told me about Chivers, how he came to the island and something about buried gold.

"Who was Chivers?" Jeanne whispered as though she might be overheard.

"A hermit, the first owner of Wallace. He planted the apple trees and lived here alone—for thirty years, I think."

We were relieved when we got outside. "Gosh," Jeanne said, "I can't imagine anyone wanting to live in there." She didn't know what her future held in store.

In the short time that remained, we decided to explore the north shore. Along the tideline we scrambled over boulders and sea-smoothed logs and across spits of powdered white shell. We studied pools of trapped seawater alive with sea urchins and darting bullheads, laughed at each other sitting in sandstone tubs hollowed out by the wind and sea. Gulls swooped and screeched overhead. Around each point, another hidden bay begged exploration. What a beachcomber's paradise! Jeanne's pockets already were sagging with bits of sea-polished wood.

Cutting inland, we descended the slope of a long meadow bounded by majestic firs. It flowed through the heart of the island like an undulating wave of brown velvet. On the south was an orchard that had been engulfed by firs. As we approached it, Jeanne spotted several gaping holes in the ground.

"Treasure hunters dug them, searching for gold," I explained, remembering now what Vic had told me. "Chivers' gold. He was a retired prospector, always paid for his purchases with goldpieces. But none were ever found after he died."

Today his romantic story still is told—how he struck it rich in the fabulous Yukon, brought his hoard of gold to the island, and retired to a hermit's life, clearing enough land to create a fertile valley and planting an orchard that still bears delicious fruit. Fortune hunters seeking old Chivers' gold have come and gone. Only the scars of the deep empty pits endure, while his buried cache of gold remains undiscovered. But, thinking back, we're sure that we discovered his treasure after all.

The sun had fallen behind the tops of the jagged firs. Tired and bedraggled, we hurried along a fir-needle path toward the cove and plodded into the old clearing, retracing our footsteps in the crumpled grass. The forest hovered about us, a mass of somber green. Under an ancient apple tree, we stole one last look at the cozy basin scooped from giant evergreens. Across the cove, fingers of light seeped through the trees and painted the lonely shack a muted red and gold.

"What a beautiful homesite," Jeanne said.

"Yes"—I nodded, surveying the old orchard—"and these fruit trees might still bear lots of fruit."

"There's good garden soil here too." Jeanne dug her toe into the ground, then walked to the edge of the cove. "With a landlocked harbor at your front door, what more could you ask?"

We hadn't noticed our own appearance. Our faces were streaked with dirt and our legs bruised and scratched from crawling through the brush. No longer did we feel or look as if we belonged to the same world we had left that morning.

Suddenly, "Dave! Dave!" Jeanne called from the beach. "Our boat's gone."

Scrambling down the bank, I anxiously scanned the cove for the cockleshell.

"There it is—heading toward the gap," Jeanne pointed.

Fanned by an offshore breeze, the boat was drifting toward the narrow passageway. We had forgotten all about such a thing as a rising tide.

"What'll we do?" wailed Jeanne.

"Swim for it." I was ripping off my shirt. "We haven't a second to spare."

Flinging off our clothes, we smacked the water. How cold it was didn't enter our heads. The boat was all that mattered. We spied it fifty yards ahead. Jeanne's arms reached it first. I dove under and brought her down, squirming in my arms. When I came up for air, "So you want a battle, eh?" I heard her yell, and my head was shoved under before I caught a breath.

We fought for the cockleshell until our lungs and strength gave out, then scrambled wearily into the boat, both too exhausted to claim a victory.

This time, we tied the painter to a fir limb. The salt water clung to the skin, and we hopped about the beach like wardancing Indians to get warm and dry. But how exhilarating it was to swim in the nude off a tree-fringed beach where the only footprints were our own. Breathing the scent of cedars and sea air, we were truly free, unhampered by the bonds of civilization. Nor did we feel ashamed or oddly out of place. For a moment we were Adam and Eve.

On a flat outcrop of sandstone still warm from the sun, we sprawled on our stomachs to dry and smoked silently, each wondering what the other was thinking. Finally, the weight of my thoughts became unbearable.

"Jeanne, do you suppose two people could make a living on an island?"

She looked up. "That idea has haunted me all day, Dave. I don't know how—but there must be a way."

Her answer filled me with optimism. "Wallace has been a secret dream of mine. I think the dream of owning an island is a kind of universal affliction. With the pace of life today, doesn't almost everyone long to escape to the peacefulness of an island?" I studied her face. "Maybe—maybe the answer lies with fellow dreamers. By sharing Wallace, maybe we could make a living here."

"That's it!" Jeanne said excitedly. "We could build summer cottages. What do you think? Wouldn't Wallace make an ideal resort for would-be Crusoes?" Then her face fell. "We're crazy! We haven't the money, even if someone plunked an island right into our laps."

"People who wait for the money, the right moment—well, when it comes it's usually too late."

"That's true, darling. But sometimes your enthusiasm runs away with you. I'm afraid you'll get hurt." She smiled. "Even if we can't afford a new car, we can dream big."

There was warmth and candor in the way she spoke. To be near Jeanne makes you feel good, and quiet inside, for you always know where you stand. Now, her eyes held mine. "Wouldn't this be a heavenly place to raise a family?"

An orange glow lit the sky. The sea was like black marble as we poked our cockleshell through the gap and scooted over the kelp beds toward Salt Spring Island. Our minds buzzed with fantastic plans—the best spot for a garden, the location of the wharf and cottage sites. We even agreed where the chicken coop should be! The thought of money never entered our heads. And, strangely, we never considered whether Wallace was for sale or what the price might be. The more we chattered about the island, praising its virtues and telling each other the many things we could do, the deeper we became involved.

"What would your mother say, Dave? Her baby boy way up in the Canadian wilds!"

Oddly enough, neither of us can remember who first said, "Let's buy the island." All we know is that the decision was reached in a rowboat in the eerie pinkness of an April twilight, somewhere in the Trincomali Channel between Wallace and Salt Spring, and loud enough for the whole world to hear.

CHAPTER 2

Burning Our Bridges

NEITHER OF US HAD THE BACKGROUND OR EXPERIENCE to make our dream remotely practical or real, let alone the money it required. We were young city-dwellers right to the core—the city was the only place we knew and understood. The closest we had come to a cow was on a billboard as we drove down Wilshire Boulevard. If something went wrong or needed attention outside our narrow sphere of know-how, we were helpless. That is, until we consulted the yellow pages of the telephone directory.

That night, drugged by the delight of our dream, we sauntered hand in hand up the rickety old wharf in the dusk, and along the dirt road until we spotted the flickering lamps of the lonely Bettiss farmhouse among the trees. Until yesterday, when we spent the night with Vic, I had not seen him since he was caretaker of the boys' camp.

Vic had grubbed a patch of land from the forest and settled back to farm. He was top man of his five acres. When he called his cows, the chickens flew into the trees and atop the sheds and his countless cats raced under the house. A bundle of excitable vitality and kindly ways, he was saturated with a "captain" complex and was always happiest when giving advice. He knew a little about everything and a whole lot about Salt Spring, including every single soul.

Now, Vic's whiskered face greeted us at the door, and his strong callused hand grasped mine. "Thought yuh kids was lost," he said as he drew us into the bright kitchen.

He was the same old grizzly rascal—hard, toothless, and unchangeable as a chunk of concrete. The stubby beard, rowdy chestnut hair, rollicking blue eyes, and boyish grin made it impossible to guess his years. Smelling of barnyard, his faded jeans hung above his ankles like a scarecrow's, and his hip pockets sagged with notepads and tools. Something about his bright eyes and the way he walked reminded me of the bantam rooster that strutted about the yard.

"Beginn' to worry about you folks," Myrt said, ushering us into the living room. She was a tiny Cornish woman, slightly grayed, with dark skin that glowed from the outdoors. She smiled at us with her soft, sparrow-like eyes. "Supper won't be long."

We dropped into the huge leather chairs around the fire. I glanced at Vic, remembering him as a caretaker. He had not been an ordinary one, for he had many worldly know-hows and a quiet, salty way with boys. His campfire yarns were the event of the week. A master storyteller, he spun a lifetime of incredible adventures—as a night orderly in an insane asylum and a bosun under a drunken sea captain aboard a threemasted schooner. At camp, if the gas lamps failed, the launch sputtered, or the cook ran short of water, the cry went out, "Get Vic! Get Vic!" Soon his thin, wiry figure, like that of some old mountain man, thumbs tucked under his braces, descended calmly into the trouble. Fingering his tobacco-stained beard, he invariably bawled, "Holy tin pots! What's all the fuss?" His philosophy was simple: "Jist do what's to be done with no belly-achin'."

I shoved Vic a cigarette. "Thanks for letting us borrow your boat."

"That's all right, my boy. How did the island look? ... Didja know it was fer sale?" A smile crossed Jeanne's face. It was as if Vic had read our minds.

"We're both crazy about it," I told him. "I had no idea it was for sale—golly, we sure would love to buy it. You know the price?"

He shook his head. "Won't take long to find out, though. You know th' Sultan, th' Ganges' storekeeper? You won his ice cream contest, didn'ja—th' time yuh ate them twenty-seven cones? He'll remember you, Tiny. Island's his now. We'll give'm a buzz after supper."

Our minds were not on Myrt's dinner that evening. I could hardly wait to talk to the storekeeper. When I did, he remembered me all right, especially the fact that I was American. He had the notion that all Americans were loaded. So was his price—to the tune of $20,000. We were sorry we phoned, for our dream burst like a soap bubble.

All at once Jeanne's expression changed. "We forgot one thing—the exchange rate. In our money, the island would be only $18,000. Doesn't that sound better?"

My spirits began to soar. I fidgeted in my pocket for a pen and scrap of paper. "Let's see...for two hundred acres...that figures out to exactly ninety dollars an acre. That's not bad, really."

Excitedly, we began enumerating all the virtues of Wallace, agreeing that it was the most enchanting two hundred acres anywhere. Though only a quarter of a mile wide, it is nearly three miles long. Its wooded contour, gently creased by one long valley, soars to 250 feet above the sea along curvaceous pink ridges of sandstone topped with oaks, junipers, and husky firs.

"If you can find them," Myrt put in, "there's pear and plum trees too. Have you ever canned fruit?"

Jeanne shook her head. We had never been around country people before. We envied them. Their earthy staunchness made us feel ashamed that we knew so little about weather, nature, shelter, and preserving food. It was as if a vital part of our lives had been neglected.

"Don't firgit," Vic added, grinning inside a cloud of cigarette smoke, "yuh got a dandy well. Must be covered by bush now—a deep bugger, too. Never did make good tea tho', did it, Myrt?"

"How about the fruit trees?" Jeanne asked. "We'd have apples to sell."

"Couldn't unload 'em around here." Vic shook his shaggy head. "Might sell 'em to them fancy boats though."

"Wouldn't the cove make a great yacht basin? We could put in floats and a gas pump and store," I suggested.

Not a ripple showed on Vic's leathery face as he reached for his tobacco can and began to roll another cigarette. The delay gave authority to his words. "That'll take a tub of money. If you ain't got plenty, son, you can't hope to git it here. Unless yer rich, banks won't lend yuh a dime."

"You mean to tell me that if I needed a couple hundred dollars, I couldn't borrow it on the island?"

"That's right, son. Jist remember—don't git yourself hung up with the Sultan."

"Never mind," Jeanne said encouragingly. "We'll latch onto something else."

"If it's income you want, run a few sheep, maybe a cow or two." Myrt frowned at him as he slipped off his shoes and stretched out his stockinged feet. "Sheep are your best bet, though," he went on. "Lambs fetch a good price too."

We agreed that with a little hard work the tumble-down shack could be fixed into livable summer quarters that would serve as a kitchen-nucleus for a house. We could add a living room, bedroom, and bath as soon as we got around to it. It would be a challenge to our ingenuity to use all the resources the island provided: small trees for beams and joists, driftwood boards for sheathing, and so on. Wartime shortages, Vic warned, had made building materials hard to get. We were lucky to have the old camp buildings, if for nothing else than to salvage nails. "They're scarcer than snake legs," he said. "Been itchin' to git my hands on enough to fix up the barn."

Jeanne spilled over with country enthusiasm. She announced quite brazenly that she could learn all the tricks of a wood stove. But it was mainly the economy of fuel that caught her attention. "Look at the free wood—the beaches and forests are full of it."

On into the night we talked great plans, pulling them to pieces and searching the wreckage for practical clues on how to buy an island we couldn't afford. Finally we wrote down our assets. Since our marriage, we had scrimped together $1,200 in War Bonds. This was the nest egg we had promised ourselves never to touch. Our pockets held $62.50 in cash. We owned no stocks or investments. Our bank balance was $492.10. My G.I. life insurance policy had a loan value of perhaps $200. Our home in Los Angeles was almost paid for. If the real estate boom lasted another month, we could sell it and pay off our mortgage, leaving us with $7,500—if we were lucky.

That left only poor old Roberta, our prewar jalopy. Even with her 165,000 miles, we would need her more than ever. So we decided not to consider her. Total: $9,454.60. Certainly not enough for us to go overboard on a costly dream.

"Darling, let's just talk to the storekeeper on the way to the ferry," Jeanne pleaded. "We'll find out his rock-bottom price and his terms, then consider an option. That'll give us time to get home and explore all the angles. . . . Wallace is my dream too," she added, with touching faith.

The next morning, the gray-haired storekeeper professed reluctance to sell Wallace. Either this was one of his tactics, or he was deeply attached to the island. We didn't know which. He stiffened at the idea of reducing the price.

"Twenty thousand, that's final." His hand slammed the oak desk. "Take it or leave it."

I stared gloomily at Jeanne. What was the sense of haggling over terms? To my astonishment, she smiled softly and asked him what the terms were.

For an hour we dickered. Not a smile came to his lips. The lowest down payment we could wangle was $6,000, with a mortgage to be paid off at a thousand dollars a year.

We stubbornly held out for more time on the first payment—a year from the coming September thirtieth—explaining that it would take us two summers to get on our feet. Pleased with our frankness, he grinned for the first time, and gave in to our demands.

We drew a deep breath and wrote a $300 check for a thirty-day option. He refused to take less or allow us a day more. We knew the chances were slim that we would ever see the money again.

In great haste we headed for Los Angeles, stopping briefly in Sidney at the Canadian Immigration Office. We wanted to be sure we could enter Canada as settlers and still retain our American citizenship. Then, driving night and day, we hurtled down the coast in a turmoil of excitement, tormented by the question: Shall we do it? Neither of us wanted to say No and neither dared say Yes. At last, as we neared the outskirts of Los Angeles, Jeanne dumped the future in my lap.

"A man's work is his life," she said. "He must love what he's doing or he can't be happy. It's only right that you should make the final decision."

Her words touched a soft spot in my vanity. Yet one doubt remained. "Would you really be happy stuck on an island, away from it all?"

Her warm hand squeezed mine. "I'll be happy wherever you are. It's up to you, Dave."

We rolled across San Fernando Valley, crept up noisy Cahuenga Boulevard, and dropped into the congested city as newsboys tore apart their bundles of evening papers. We had almost forgotten how our hometown looked and how much we had overlooked. At Hollywood and Vine, watching the surging mass of humanity around us, pushing and rushing, homeward bound, a strange depression fell over me. For most people, I wondered, had this been just another day sold for another dollar? One more day empty of living? Every time Roberta stopped, I found myself staring at the crowds congregating at the traffic lights—tired, worried people caught in a relentless chase, prodded on blindly, driven by fear. It was written on their pinched, somber faces. Fear of the clock, of mistakes, of a fall in the market, of the weather, of what others think, of losing their jobs, of what tomorrow would bring—every step and desire shackled by fear. In the multitude of faces sweeping by, I saw the sadness of buried dreams—so few heads held high.

The island loomed larger in my thoughts. Now it became something more than a boyhood dream or a city man's escape to a simple life. The island was a challenge, a challenge to know myself, the real self hidden beneath my fears. The escape was here, all about me. I remembered Haig Brown's words, "City people talk of escape when they talk of country living. To me, a cow is reality. Escape is homogenized, pastuerized milk, delivered to the back door in a disposable carton."

The shallowness of our lives filled me with shame: the movies, the "hot spots," the flashy cars, the super-drugstores; packing our unpaid-for homes with still more unpaid-for gadgets; the endless pursuit of excitement to protect us from boredom. Yet, in reality, were there not talents asleep in each of us that only needed awakening? What a crime to keep them buried, chained by fear and conformity. I looked down at my broad hands gripping the wheel. Surely they could do more than jockey a car, trip a camera shutter, write checks, push a cart around a supermarket? Suddenly I felt as if I had not tasted life. The joy of self-discovery still awaited me.

As the street lights flickered on, Roberta rubbed the curb in front of our pink stucco house. We were more relieved than happy to be home. Avoiding serious talk about the island, we nibbled our dinner with little enthusiasm, bathed, and crawled into bed. Exhausted as we were, it was impossible to sleep. My head throbbed as I pondered what course to take.

There was little chance of raising more capital. Our friends were mostly aircraft workers like ourselves. We had no fathers. Jeanne's mother worked in a library; mine lived on savings handed down from Dad. Had I the right to ask Jeanne to give up her comfortable existence for the hardships and doubtful security of island life? Both of us had been happy and satisfied with our lot until now.

"Damn it all!" I shot up in bed, reaching for the light. "Let's tackle it. What do you say?" I hovered over Jeanne.

She threw her arms around my neck. "Why not? We're young, healthy, and strong. The chance may never come again."

Far into the night, we excitedly combed over our plans, christening our adventure over a quart of chocolate ice cream.

The next morning I awoke with a devilish urge. While Jeanne was still asleep, I slipped quietly into the kitchen. Sure enough, old lady Crawford was puttering with her breakfast. I glanced guiltily back at the bedroom door, then stared out the window until her baleful eyes caught mine. Wrenching my face into an ugly contortion, I slammed down the blind.

"My, you look smug this morning," Jeanne commented at breakfast. "I'm glad you do. You'll have to face your mother alone. Blood is sure to fly—and it's not going to be mine. Anyway, her face drops a mile when I tag along. She never—"

"Ridiculous!" I cut in. "What mother thinks her son has married the right girl?"

"But it's true. Sometimes our natures are so different it frightens me. I'm not the calm, collected soul you are. Even your patience exasperates me. Why, I've seen you spend hours stalking a bumblebee to get one camera shot."

"Listen, my sweet, if you were like me—a pigheaded, fussy, stiff-jawed, impractical introvert, I couldn't live under the same roof with you. Besides, some authorities say opposites make the best marriages."

Our friends soon heard about our plans. When we first explained our intentions, they just laughed as if we were not in earnest. But after the FOR SALE sign was planted on our doorstep and we were seen shopping at Sears for a car trailer, they realized we were serious—and shuddered and flung up their hands: The Conovers had gone berserk.

Our excitement grew as each day passed. We found ourselves drinking coffee at all hours, wondering when the house would sell, planning what to take and what to leave behind. As prospective Crusoes, we raided the library for books about island life and boats. We wanted a powerboat that could be easily handled by one person, yet was safe enough to weather the roughest sea. Our budget dictated that it must be a small open skiff, and a secondhand one to boot.

Every day the phone jangled with warnings mixed with advice. "What's this I hear about you making tracks to British Columbia?" I overheard a girl friend ask Jeanne. "To an island, of all places! You aren't the type!" Jeanne's office boss minced no words: "You'll either freeze to death or go off your rocker!" An old school chum, Bill Harvey, was clearly irritated at me. "Good lord, man—bury that pretty gal on an island! You'd better see a couch doctor in a hurry."

Oddly, the only note of joy came from my mother-in-law. Alice confirmed my conviction that I had two things in life to be grateful for: a loving wife and a loving mother-in-law. Seldom is a man blessed with both. She is one of those wide-eyed, peppy little women who hide their true age behind a cheery disposition and a venturesome heart. On every birthday she pegs her age at forty, then spends the rest of the year living up to it.

When I told her I was going to take her daughter off to an island in the Canadian wilds, her face lit up. "Could I come along?" Seeing my shocked expression, she went on quickly, "Don't worry— you'll have your hands full enough with Jeanne."

It was nice to know someone approved of our dream.

A week flew by before I gathered the courage to climb the steps to Mother's apartment. I spilled out our plans in a torrent of enthusiasm.

She erupted like a volcano: "You're no Robinson Crusoe! How in thunder would you make a living on an island? Why—the whole thing is fantastic!"

I was stunned by the fierceness of her disapproval.

"Young man, look at yourself." Through her horn-rimmed glasses she glared at me as if I had committed a crime. "You may be big, but you're soft—you're no pioneer. Why, your skin is so sensitive it's allergic to Lux. And look at your hands—they've never known a callus. The only tool they've ever touched is the screwdriver that came with an Erector set. You're a dreamer, David—the artistic type, not a roughneck adventurer!"

The best thing to do, I decided, was ride out the storm in silence.

"What's more, young man, you don't have the money for such a hare-brained scheme. You're a darned good photographer," she added proudly, a note of sadness creeping into her voice. "Are you willing to throw your career down the drain?"

"I'm sorry, Mother," I said, to avoid further argument. "My mind is made up. We'll be leaving sometime next week."

Mother never broached the subject again.

Because of the postwar housing shortage, selling our house was easier than we expected. A week before our option on Wallace expired we made the down payment and signed the $14,000 mortgage, not seriously concerned about this fateful document we were so lightly scratching our names to. By mail we became the proud if somewhat encumbered owners of our own private kingdom.

With the deed to paradise clutched in our hands and our house promised in two weeks, the mad scramble was on. Jeanne proclaimed May tenth Departure Day, just thirteen days away. There were a thousand and one things to be done—all at once, it seemed. For two spoiled Californians to trade their suntans, smog, and supermarkets for a remote Canadian island with no house, no electricity, no telephone, no running water, not even a boat to reach it—well, it was like preparing to migrate to the moon. We hardly knew where to begin.

We made lists, tore them up, started new ones. We sweated over a budget, but time after time the imponderables bogged us down. What was the cost of a boat, of building materials in a strange country? How far would our capital stretch? Were we overly optimistic in counting on a small income—a cottage finished, ready to rent—next year? How little a month could we possibly live on?

Our heads were in the clouds. We collided in the hallway, ran the bath water over the tub, stumbled over piles of clothing and the boxes and crates that cluttered every room. One morning I backed out of the garage to the ear-splitting sound of splintering wood, to find I had forgotten to raise the garage door. The next night we locked ourselves out. Under the curious, puzzled eyes of our neighbors, we stood on the garbage can and forced entry through a high bedroom window with a shovel blade. We were sure they expected the "padded" wagon to screech up in front of our door at any moment.

Another day, we stood knee-deep among our possessions, wondering what to do next. "Let's tackle the shopping list," I suggested. "And pick up the trailer."

The luggage trailer looked huge compared to all the others on the lot, but when it was bolted onto Roberta, it seemed to diminish in size—particularly when we lurched away from Sears with it crammed full of shovels, sleeping bags, gas lanterns, candles, a fuel can, tools, tent, unpainted cupboards, long johns, boots, mackintoshes, blankets, and a rural mailbox. That night we had to revamp our budget.

We agreed to strip ourselves to the bare essentials and squeeze everything into our rolling stock. As if we were preparing a small aircraft for a long ocean hop, we had to consider limitations of both space and weight. Every item not only had to be practical but must pay its way in usefulness. The countless decisions were heartbreaking.

"Good lord! We can't take that," I cried, pointing to the hall mirror.

"Why not? It's our loveliest wedding gift."

"Darling"—I tried to remain calm—"what on earth is practical about a six-foot plate-glass mirror? Besides, it's heavier than lead."

In the end, neither of us gave in on two articles that we considered essential: Jeanne's old Singer and my father's red mahogany desk.

When she caught me wrapping the desk with her rugs, she protested, "You're not taking that monstrosity! Looks like the desk was thrown into the bargain when we got married."

"Then I guess you won't need your sewing machine, will you?"

The discussion ended right there.

To fill our coffers, Jeanne sacrificed her new automatic washer, vacuum cleaner, steam iron, pop-up toaster, waffle iron, and—not without a last-ditch stand—the Mixmaster, her pride. "It won't take much room," she begged, her eyes hopeful. "Can't we take it?"

It was hard to refuse her. "We made a rule, remember? Let's stick by it."

D Day was only three short days away when we began the long tedious task of loading all that remained of our belongings into a five-by-seven wooden box on wheels, reserving Roberta's turtleback, which we could lock, for traveling gear, food, and suitcases. We sweated, puffed, and pried until, late the following day, the job was done.

"Toss this canvas over the top." I handed one end to Jeanne. We had just started to lash it down when we noticed the tires. They were almost flat. I reached for the pump with a silent prayer that they would hold.

That night Ralph and Dottie, our dearest friends, finally showed up at the house. We eagerly whipped out our deed to show it to them. Of all our friends, we had counted most on them for good wishes or at least sympathetic understanding. We were dismayed at their reaction.

Dottie reeled out all her heavy artillery. How did we expect to start a resort—we knew nothing about one—on an island? We hadn't that kind of capital, and we'd kill ourselves earning a living. How could I build a wharf, water reservoirs, roads, and cottages when I'd never sawed or hammered a board? Neither of us had owned a boat or even knew how one worked. We might get seasick. One of us might drown and leave the other marooned to slowly starve to death. And if that didn't happen, then the isolation with its enforced intimacy would certainly drive us at each other's throats.

In a way, all these objections were helpful. We had convinced ourselves that to know the problems we were going to face was half the battle. Already stowed in cardboard boxes were books about boats, seamanship, engine mechanics, cabins in the woods, vegetable gardening, wood finishing, chicken raising, forestry, interior decoration, and carpentry, as well as plumbing and mason's guides (8 volumes), *How to Operate a Small Business, the Complete Home Doctor, Sheep Husbandry, 100 Ways to Prepare Sea Food, How to Prune Fruit Trees,* and *Expectant Motherhood.* Jeanne laughed. "Now we're prepared for anything."

After Ralph and Dottie were gone, I lay awake thinking about what he had said to me. My heart was heavy for him, though he had much to envy. A successful junior executive, a kind, lovable man living for his wife and children, to all appearances he seemed happy and contented. However, that evening he had sat on a packing crate with a far-off gleam in his eyes, nodding token approvals while Dottie, upset by our foolhardiness, ruled the floor. When they were ready to leave, Ralph motioned me to step outside and walk with him to the car.

"Don't let Dottie discourage you, Dave," he said as the front door closed behind us. "I wish I was in your shoes. You know, I don't have the guts to break away."

I knew the dream that he held in his heart. He had told me about it years ago, the day we ditched school together and sprawled on a sand dune, the sea air whipping our faces. His silence, now, told me he still cherished that dream.

Within every man there lies a dream. It is to his mind what his heart is to his body. He not only believes in it, but he lives because of it. He nourishes it with thought, unaware it is his most valuable possession. It sustains him in times of crisis. It comforts his fears and silently molds his future. Wherever he goes and whatever he does, it remains part and parcel of his being. And when his dream is fulfilled, he quickly plants another. For without a dream, a man may still exist but he ceases to live.

Our dream had persistent foes. We knew our friends scoffed at our stupidity and were chanting "You'll be sorry" right up to the last day. But we also knew that their clamor of disapproval, more strongly voiced by the women than the men, revealed a deep concern for our welfare, for with touching affection those who had protested the loudest came at the last moment bearing farewell gifts. We mumbled words of thanks for life preservers, seasick pills, a first-aid kit, canned butter, flashlights, and books. Nearly always, they departed with more than they had brought: picture frames, records, flowerpots, coat hangers, and doormats.

Oddly, the day before our departure, the most frightening words of all came from the carpenter who was repairing the garage door.

"I hate to alarm you, Mr. Conover," he said in a low voice as he measured a slab of plywood, "but a friend of mine who takes a Vancouver paper told me three white people were murdered recently by Indians." He looked up. "Do you know where Salt Spring Island is?"

There was an Indian reservation, I remembered, not far from Wallace. The story might be true. However, such a tragedy was probably no more than a rare, isolated incident, and I thought it best not to mention it to Jeanne.

For days we had been far too busy and excited to consider our true feelings. Now, suddenly, all the pitfalls of our adventure having been exposed and duly ignored in the hubbub, the first pang of remorse assailed us. Had we mistakenly bargained for a rich man's toy? We knew that to survive as islanders we would have to become sailors and master the sea in a small open boat. Were we so wise, after all, to gamble our lives on a dream?

At last the big morning came. We had announced nine as the hour of takeoff, and well before then our families and friends were milling around in front of the house, waiting to say good-bye. Their faces were solemn and tense. It looked like a gloomy gathering of mourners in front of a funeral parlor. We rushed about shaking hands, inviting people to visit us, trying to look pleasant and bring a quick end to the ordeal. But everyone kept standing and talking, ignoring the signs of our impending departure. Finally, I leaned on Roberta's running board and blew the horn. The huddles began to dissolve. One by one, in slow motion, the "mourners" shuffled forward in a straggly line—eyeing our caravan dubiously, cracking painful bon-voyage jokes, holding out clammy hands to wish us luck as if they never expected to see us again.

The one gay note in the festivity was the last-minute arrival of Bill Harvey, who had forsaken his Hollywood route and driven his Good Humor truck hell-bent across town to pull up beside Roberta, bells a-jingling. His huge white hulk bounced from the truck. Beaming like a jolly Santa Claus, he dispensed ice cream bars to his astonished audience. Smiles at last began to spread over the worried faces. Big Bill had saved the day! At the very end, the walls of reserve crumbled between our mothers, and in a burst of tears, they fell into each other's arms.

"Let's go before I start gushing too," Jeanne sputtered from behind a Kleenex as she climbed hurriedly into Roberta.

We took off to a mad jingle of chimes and bells and a flurry of shouted farewells, Roberta weaving and swaying down the street in response to the pull of her trailer offspring. As we turned the corner no more than six blocks from home, there was an horrendous screech. Roberta squealed to a burnt-rubber stop.

"There go the springs," I groaned, so exasperated I was tempted to pull the steering wheel out by the roots.

We sat in the car and moped until we were sure everyone had gone, then sneaked back home. That afternoon we had heavy-duty springs installed. The next morning at daybreak we crept from the house and headed north on Highway 99.

Roberta huffed and puffed up the grapevine. A flat near Weed—the middle of nowhere. The damnable Siskiyous! Lost in Portland. "Will it ever stop raining?" Puget Sound—gateway to paradise. We began to relax. Port Angeles—a pimple on a peninsula. At last, Roberta took a ferry ride.

CHAPTER 3

The Desperate Voyage

ON A MAP OF BRITISH COLUMBIA, DRAW A LINE LINKING the cities of Vancouver, Nanaimo, and Victoria. In the heart of this triangle lies Wallace Island, nestled in a little-known group locally tagged somewhat paradoxically as the Gulf Islands, since there are no gulfs anywhere in British Columbia. So, taking a cue from the famous ships that ply their waters, we soon found ourselves calling them the Princess Islands.

Unspoiled by civilization's creeping hand, the islands rise from the sea cloaked in rich hues of brown and green, circling the southeast shores of huge Vancouver Island like a necklace of gems. Over a hundred in all, they range in size from Salt Spring Island's seventy square miles to rocky islets barely large enough to pitch a tent on. Their names read like a history of intrepid Spanish and English explorers—Galiano, Valdez, Mayne, and Gabriola. They stretch in clusters south of the forty-ninth parallel, interlocking like a jigsaw puzzle with the American San Juan Islands, forming an unseen boundary that jogs through a great inland sea of islands toward the Straits of Juan de Fuca.

Wild and sparsely populated, the islands have few settlers. Their houses hug the forested shores of larger islands, where they farm, log, or are simply retired—often waiting for checks from England. What might have been called "civilization" a generation ago blooms at its peak on Salt Spring around the sleepy village of Ganges. Here, rugged Canadians—loggers and prairie farmers—wrestle for control of the island with a colony of Britishers—majors, colonels, and remittance men. Among the latter, teatime is not only a punctual ritual of Sterling and Wedgwood, but a clannish seminar where they nurse and groom their untainted accents and discuss the diet or whereabouts of the Queen. Next to strangers, nothing is looked upon with more suspicion than change. With a peculiar indifference toward progress, the islands remain rooted to the past—static and green and nearly forgotten, sprinkled with tiny vestiges of a bygone British era.

Geologists say the islands were formed by retreating glaciers and floods during the last stages of the Ice Age. But an old Indian legend gives a more colorful explanation: Untold centuries ago, along the coast, the king of the eagles flew over an Indian village one day. Small boys began shooting arrows at the circling bird. Instead of being frightened, the giant eagle continued to circle over the village, attracting older warriors, who joined the shooting. But the eagle had superior powers. The arrows glanced off his sleek feathered body, and he continued to circle defiantly. Soon, all the chieftains and warriors of the tribe were whistling arrows from their bows. Suddenly the bird swooped down over the compact group and seized one of the men by the hair. A second warrior clutched at the legs of the first, but the eagle soared upward. One after the other, the men and boys of the village clutched and grabbed at a neighbor's legs—until all were drawn into the sky. They could not let go. Straight for the sun flew the eagle, with the human chain attached. When the whole coastline could be seen below, the bird began to swing its human cargo in ever widening arcs. At last it released its mysterious hold, and the Indians plummeted into the sea. Where each fell, an island arose, the rank of the brave determining its size.

Soaring above the islands like a paternal watchdog, the bald eagle is still master of the skies today. He circles the Indian villages, watching their wooden dugouts as they fish the tide rips for salmon. Nothing escapes his vigilance. He watches mink searching for crabs along drift-strewn shorelines; black cormorants drying their wings on sea-washed rocks; ducks swishing to rest in secluded coves, and white-tailed deer roaming the underbrush by day, swimming from island to island by night.

But now, above the largest island, he also sees farms being etched from the forests, and men bringing in hay from lonely fields. He hears the chattering staccato of chain saws break the eternal silence as man's wicked machines claw the virgin forests, leaving a wake of bare earth. He watches ferryboats paddling like white ducks from shore to shore, and hungrily eyes seiners hoisting their nets to harvest the sea.

Here, indeed, was a sea of forested Edens beckoning us. But, unfortunately, man does not conquer paradise easily. Nor does survival alone fashion a dream. We were wrong to think that by taming the island—subduing its wilds and planting a bit of civilization on its shores—we would win our major battle. We had not foreseen, nor were we prepared to face, the greatest struggle of all: what the island would do to us.

In Victoria, we began at once to shop for a boat. The necessity of choosing exactly the right one weighed heavily on my mind. Boats were scarce, and our budget allowed only five hundred dollars. At Ye Olde British Fish and Chips Shoppe, we eagerly combed through the classified ads as the brownish lumps of fish grew cold on our plates. Only a few fishboats were listed. Discouraged, we were ready to try our last hope, the telephone directory, when Jeanne said excitedly, "Here we are! This sounds good." She pushed the paper under my nose.

USED LIFEBOAT FOR SALE—16 ft., with 3 hp. Wisconsin.
Good shape, $450. See Mr. Goodwin, Fisherman's Wharf.

Mr. Goodwin was a burly, seasoned fisherman who bore a striking resemblance to Charles Laughton. He smiled and gestured toward the boat. "Step in and look her over."

"Why it's just a king-size rowboat!" Jeanne gasped.

For a sixteen-footer, she was indeed chunky—fully six feet across the beam, but compared to the trollers and seiners around her, woefully inadequate for our purpose. Yet her deep body and lapstreak lines hugged the water with a grace and carriage that told us she was born for the sea. She was oddly different from the double-ended lifeboats we had seen on board ship coming over. She had a square stern, and a rudder with a tiller as thick and long as a man's arm. Under the gunnels wide oak seats circled the hull, two crossing amidship, one on each side of the motor. As we scoured every plank and rib, we began to sense her pride and dignity. She had a haughty air, a scornfulness that seemed to say, "I'll take a lot more than those big boys will!"

In contrast to the boat, the motor was so frightfully puny that I wondered how it could push her. "Mind showing me how she works?" I asked.

Mr. Goodwin flicked a tiny lever, adjusted what looked like a throttle control, and pulled the rope. The motor burst into an astonishing roar, and we grabbed the gunnel as the boat tugged and thrashed at the float. Any moment, I thought, the little brute would explode. The exhaust shot out of a rusty pipe like a pounding bass drum.

I stared at the motor, my respect for it sky-rocketing.

"Simple, isn't it?" Mr. Goodwin hollered in my ear.

I wasn't too sure, but I nodded.

We felt certain she was seaworthy. But was she, we wondered, the right boat for us? For the enormous loads she would have to carry, it might be like asking a child to do a man's work. Was she really big enough to colonize an island? Big enough, the year around, to ferry over heavy loads of lumber, bricks, cement, and other materials to build our kingdom?

A trial run about the harbor convinced us. Then Mr. Goodwin began to unreel the intricacies of her operation. He pried up several green floorboards. "There's no clutch," he announced in a tone that sounded as if there should be one. "You see here"—he motioned with his finger—"the shaft couples right to the motor. You must be careful when you start out," he added sternly.

"Why?" I hated to ask.

"Why, if the boat isn't tied, she'll take right off—smash into something."

I consoled myself by thinking it was better to feel silly now than sorry later.

He recommended the anchor, chain, and rope to buy and advised carefully checking the oil level, grease cups, filter bowl, breather cap, points, plug, propeller, and shaft coupling.

"Remember," he warned in conclusion, holding the starter rope in his hand, "don't forget to wind this clockwise. Else the motor won't start."

I broke out in a cold sweat. Motors and I had always ignored each other—we were total strangers. For years, mechanics had greeted me with glad hands. I wondered if Roberta really needed the new plugs they put in. And when she got new rings, whether a valve grind would have done the job instead. Now it was different. The sea would be my highway and a boat would be my car. I could no longer hide behind my ignorance.

When we were about ready to leave, I handed Mr. Goodwin the check and casually mentioned that I would be heading for the islands in the boat in the morning.

His eyes grew serious. "This is no time to travel in a small boat," he said.

"Why not?"

"You can't rely on the weather—we're still in the shadow of the equinox. Besides, you said yourself that you've never handled a boat before. And you don't know these waters." He lit his pipe and then went on gravely, "More boats are lost this time of year than any other. Better wait another week or two, at least."

With so much to be done before winter, every day was precious. "There must be calm days," I argued, trying to convince myself. "Wouldn't I be within sight of land all the way?"

"Makes no difference," he said sharply. "Squalls can strike without warning. The coast is rocky, and there are no harbors. And the lifeboat does only five knots. If you're caught in a blow, the chances are you'd never reach shore alive. Take my advice, young man—wait."

Jeanne looked at me wistfully. "This equinox or whatever it is"—she tried to sound cheerful—"perhaps, you'd better wait."

We had several times plotted on a map the prospective boat routes from Victoria to Fernwood, where Jeanne was going to meet me. The old wharf was to be our base of operations, the nearest bite of land to Wallace that could be reached by car. The shortest water route was forty-eight miles of weaving and twisting through the islands. One stretch had worried us particularly. It was the forbidding open sea at the very beginning.

Mr. Goodwin confirmed our suspicions. A dangerous area did lurk outside Victoria and for a considerable distance northward into the islands. Once that stretch was hurdled, I reasoned, the worst would be over. With an early start, I could reach the shelter of the islands by noon and, if luck held, arrive at Fernwood by dinnertime. Everything would depend on a calm sea those first few critical hours. Was this pushing our luck, I wondered?

We said nothing more about the trip and shook off our pessimism by outfitting the boat. Thanks to Mr. Goodwin's tips, we knew exactly what to buy. As the sky commenced to darken, we stowed away gasoline, oil, anchor rope, bucket, bilge pump, tools, and chart. Life jackets and first-aid kit were tucked under the transom seat, and the gasoline and heavier articles squeezed into the bow. From under the seats we pulled out the oars, amazed at their clumsiness and flagpole length.

"It'll take two people to row this boat if the motor ever conks out," I commented.

After inspecting the canvas engine-cover for leaks, filling the gas tank, and checking the oil, we had only one job left—a name. The first ones we considered were too feminine and delicate and seemed not to fit.

"How about Bertha?" Jeanne proposed. "She's not beautiful, but she's certainly husky." We were both satisfied.

It was almost dark. We sat on the edge of the float and gazed at Bertha nuzzling our feet like a baby whale. She was just a bundle of planks, a few copper rivets, and a handful of screws; yet she had a hearty personality. She gave off a feeling of strength and trust, as though she could tackle the roughest sea and the meanest task. Bertha was more than a boat—she was our new partner.

A wave of confidence ran through me. With a gallant new partner and a new world to conquer, why should I wait? "That damned equinox!" I fumed to myself. I could feel my jaw stiffen, my Dutch blood begin to rise.

"What's eating you?" Jeanne said.

"I've made up my mind. I'm shoving off in the morning."

"I thought you'd say that. Your chin has that stubborn look. Why did I have to marry a Dutchman!"

The next morning, Bertha greeted the daybreak with a gusty roar. I had talked brave, but now I squirmed inside.

"Dave, be careful—don't take any chances." Jeanne's eyes sought reassurance from mine. "Remember, darling, I'm counting on you."

I jumped into the churning boat and cast off. "See you at Vic's for dinner," I called back to her.

Bertha darted toward the outer harbor under a clear and promising sky. Off to port, the tall grain elevators of Ogden Point quickly slipped by. Then we rounded the massive breakwater and nosed northward along the jetty. I was alone, wide-eyed at the huge expanse of sea. Calm and peaceful as it was, the sight of the open straits—empty and cold green—made me feel like a child the first day at a new school. I was just plain scared.

There was too much to learn quickly. How would Bertha do in a storm? Did I have enough gas? What if the motor stopped—what if I hit a reef? I eyed the boulder-strewn shoreline ahead, the gentle sea smashing white curves against jagged miles of cliffs. Mr. Goodwin was right. There wasn't a chance to get safely ashore. This must be how it felt to fly solo for the first time—to be at the helm of your own destiny. My grip tightened on the rudder. "You asked for it, boy," I reminded myself. "You're on your own."

Sitting at the tiller, I pulled out the chart. We were a half-mile off Clover Point. For the next twenty miles along Vancouver Island, keeping well offshore, I had no serious need for a chart. Later on, I would be lost without it, picking my way through the islands to Fernwood.

Trial Island cropped up across the bow, exposed and desolate—the first of two scraggly islands off Victoria's rocky peninsula. I longed, almost superstitiously, to get past it because of its name. The sun, now my companion, climbed above the distant American Islands and painted the Olympic slopes in purples and pinks. Bertha droned on as I listened to the sea swish under her keel and let the sharp salt air relax my grip on the tiller.

By eight o'clock we were a stone's throw off the island. A wrecked fishboat was driven high on the rocks, its grayed skeleton a silent sentinel of treacherous shores. Suddenly Bertha gave a violent lurch. I dropped at the tiller to steady her. For no apparent reason she jerked sideways, then shot ahead. Now she swung around, pointing in a different direction than I steered. I clung to the tiller, unable to believe what was happening. Swaying as if she were drunk, Bertha nosed out to sea, then lurched and raced toward shore barely a hundred yards away. The tiller was powerless.

I tried to keep calm. But there it was again, more violent this time—an awful lurch as if some sea monster were playing tricks on me. Frantically I gripped the tiller, praying it would respond. Sweat was sprouting on my forehead. Next time, the boat spun in a circle and jolted to a stop as she shifted directions. The motor raced, then labored. We jerked one way, pitched another; like a wild ride in an amusement park, each spin more frightful than the one before. I peered warily over the side to see what was attacking us.

Whirlpools! All of forty feet across, they churned the glassy surface, whipping into foam at their centers as they twisted and curled under the bow. The sea trembled in mad tumult. Furious rows of wavelets, tossing themselves into sudsy peaks, attacked the hull from all angles.

These were the dangerous tide rips I had heard about. I felt one claw at the keel. For a moment the water quieted in a dark pool; then, like thick molasses, it commenced to churn and the boat began to circle its outer fringe. Ten feet off the bow a black hole deepened before filling with swirling froth. The rudder was hard over. We spun, lunging toward shore—stern first one second, bow first the next. My breath caught in my lungs. Fifty feet away now, waves boiled over the kelp-stained rocks. I gaped in terror.

The open sea lay flat and glistening off the north tip of the island several hundred yards away. When we swung and momentarily pointed to it, I slammed the throttle. Jostled and swirled, we charged through the slop until we were free. Weak with relief, I shakily lit a cigarette and listened to Bertha purr sweetly again.

It was ten o'clock. After bathing my face in seawater, I took a fresh look around. How strange—still not a ship in sight. Ragged wisps of clouds scurried across the sky as if to seek the safety of the horizon. To the east, snow-coned Mount Baker stared down with cold aloofness. And coming closer every moment, Discovery Island—wooded and plump—lay before me. The chart showed a tricky passage between it and the shore. To save a couple of miles, however, I wasn't going to chance it. I headed Bertha around the island as Haro Strait yawned through the gap.

Beginning where the Strait of Juan de Fuca leaves off, Haro Strait sweeps northward along the Saanich Peninsula like a broad alley, and melts into the Canadian and American islands, splitting into channels, passages, and sounds. Studying the chart again, I decided on a course straight down the center. My first lesson had taught me the false security of sticking too close to shore.

Shortly after eleven Bertha nosed into Haro, bucking a brisk north wind. Miles ahead, the Princess Islands loomed on the horizon like a scene from a technicolor travelogue. They popped up like hills of mushrooms, their green feathered tips flattened by the scurrying clouds. Only by the infinite hues of green could one tell them apart.

For a sight down the glistening strait, I aimed Bertha at D'Arcy Island, the greenest dab of land, eight miles away. Before reaching the treacherous shoals known as Kelp Reefs off its southeast tip, I planned to veer off the strait and run up Sidney Channel into the islands. Once past D'Arcy and under the lee of sprawling Sidney Island, I figured, the sea would be the least of my worries. From then on, it was only a matter of navigation, relying on the chart to identify islands and avoid submerged rocks, to weave my way to Fernwood.

Noon came and went. I stifled my appetite, hoping to celebrate lunch in the islands. Bertha droned on, nibbling away at the last stretch of open water. The tenseness began to wear off. Enervated by the brisk air, I relaxed my grip on the tiller and rested against the transom to drink in the wonder of sea and islands. It was as though all the windows of my mind were flung open.

What a change this was from crossing on a Princess ship. From its decks, you feel the sea as an impersonal thing, as impersonal as the landscape you view from a speeding car. You are never part of it. On decks of steel you are safe, surrounded by a bit of man-made land set adrift. But put to sea in a small open boat, and you meet it intimately. Every sense springs to life, vibrant and tingling. Drop your hand over the gunnel, your fingers touch the icy water. You hear and feel it slap the bow, rumble under the keel, and gurgle behind the stern. When the salt spray bites your tongue and stings your cheeks, and the salt air fills your nostrils, pure and sharp, undiluted by distance—you're as close to being at sea as you'll ever come.

Now I realized that in all my twenty-seven years I had been like a passenger on board ship. I had seen much, but knew little; done much, but lived little. That was past. Ahead lay fascinating horizons to explore, for an island is a ship always at sea. With a lifeboat linking me to the outer world, what more could I ask? Only, perhaps, a sailor's knowledgeable ways of the sea.

In spite of my fears, I could feel the sea creeping into my blood. Its secrets intrigued me as a boy is bewitched by a sailing ship imprisoned in a bottle. Once I got the knack of handling Bertha and proved my worth as her captain, I knew my fears would dissolve. Good captains don't lose good ships, and Bertha was a rugged sailor. She could take it. The question was: Could I?

By early afternoon, with D'Arcy Island still three miles away, the sky told me what I was in for. From the east, dirty black clouds began blotting out the sun, scudding low and menacing over the islands, hiding all but the nearest from view. Swinging to the south, the wind stiffened, churning the sea into tumbling whitecaps and billowing my jacket out like a sail. Bertha's exhaust grumbled as the sea swept under her. I reached for the life jacket, welcoming its warmth around my chest. The waves were higher now, shooting the boat ahead, then tugging at her stern as if not to let her get away. To keep from shivering, I squeezed myself against the tiller and pulled out my last cigarette.

Shortly after two o'clock, my most persistent fear began to take shape. What at first had seemed only a passing squall became a raging gale. The wind tore at my clothing, howling as it exploded into rain, which stung my hands and riddled the water like buckshot. Hurriedly, I tossed the canvas over the motor and slumped uneasily at the tiller, watching the gray dim world draw around me.

Shrouded in rain mist, D'Arcy Island stood off the bow the distance of perhaps two city blocks away, the only land visible to my straining eyes. Mountainous waves lashed its rocky shores feather-white and leaped up hungrily at its greenery. This lonely scrag of land was all that I had to go by now. I couldn't afford to lose it. Yet I was as close to it as I dared come. Off to the starboard—somewhere—were the Kelp Reefs, acres of rock that pockmarked the chart like submerged mountaintops. To give them a healthy clearance, I swung Bertha toward Sidney Channel, sliding down rumbling combers that followed her stern like a pack of white wolves.

I glanced at the soaked chart. In a few minutes now, I knew, I'd be under the shelter of the islands.

All at once, the motor sputtered and konked out, and Bertha began to roll and pitch savagely, broadside to the seas. Heart pounding, I clung my way to the motor. The gas tank was empty! A great roller rammed the starboard rail just as I reached for the gas can, and the next thing I knew I was sprawled on my back staring up at the sky, soaked and aching all over. Fortunately, the life jacket had broken the impact. When I began to pull myself up, I noticed gas from the overturned can dribbling into the bilge.

Another wave hurtled the rail as I lunged for the can. Thank God, it was still half full. Shivering with cold and fright, I wondered if there was enough. But the main thing right now was to get the engine going again before Bertha was swamped.

With a fretful eye to sea, I filled the tank and flung the canvas over the motor just in time. For I saw it coming, rearing up out of the murk fifty feet off—a mountainous foam-crested wave far larger than the rest. Terrified, I clung to the seat thinking that surely this was the end.

Whoosh! For an instant the world went dark, upside down, and deathly cold. Crushed to the floor by an avalanche of boiling sea, I blinked open salt-stung eyes, my head throbbing with pain. Water sloshing against my limbs reminded me of my plight. I sat up dazed, realizing the sea had saved me only momentarily and was toying with my life.

Another great wave slammed the starboard. Bertha shuddered in protest, tossed and pitched in crazy abandon. Each time the rails dipped, I thought she'd go under. Water swirled about my waist, alive with floorboards, oars, gas can, life jacket, and bucket. The clang of metal unnerved me. What should I—could I—do? Was this what happens when a wild dream runs away with you? Not a trace was ever found of Halliburton, Fawcett, or Slocum. Why, for God's sake, had I let a dream take over my life?

I felt dizzy and weak, and wanted to vomit but couldn't. Shivering, I staggered to my knees, my numbed hands fighting to hold onto the gunnels. The wind plastered the wet suntans to my skin like icepacks as I searched the howling grayness for some sign of a ship. My heart sank. But I kept on scanning the rumbling seas, blinking the salt from my eyes, praying for help. All the time I could see the rollers steepening, looming up like overhanging cliffs, the wind whipping the snow from their crests. Icy gusts shrieked with a vengeful whine. There was mockery in the wind. "So you want to be a sailor, do you?"

Any moment, I was afraid, another giant roller would raise its monstrous head. I had read somewhere about these huge waves, how during terrible storms they came in frequent cycles. But even in my wildest dreams, I had never expected that I would spend my last moments waiting for one. I felt like a prisoner on the verge of execution. My mouth was dry, and I craved tobacco.

I looked down at the canvas-covered motor, lashed now by the havoc in the bilge. It still might go, but how could I stand up to crank it? And even if I succeeded, once underway the shifting bilgewater would swamp her stern. But I groped for the bilge pump and began to pump for my very life. Then, discouraged, I flung the pump angrily back into the bilge—I couldn't compete with the sea that was surging the rails. The water was slapping higher about the motor. In another minute it would be too late. Desperately, I seized the bucket and fell to bailing. Again and again, I bent and scooped until my arms and shoulders were numb with pain, but it was like bailing with a thimble.

For an eternity Bertha wallowed helplessly in the troughs, her buoyancy gone, battered by wave after wave, while I clung to her pitching rails wondering how much longer she would stay afloat. As the minutes of misery dragged on, I drifted into a coma of indifference. Nothing seemed to matter now. Even God had deserted me.

I had never faced death before. Now that I did, I began loathing myself. For I knew I was a coward. I was giving up, waiting for the end without a shred of guts—a traitor to my ship.

Furiously, my mind recoiled. Where was my faith? If a man must die, he should go down fighting for his ship, his life, for what he believes. Otherwise, living is meaningless and futile. We are only as strong as our faith. What about Jeanne? What would happen to her? She had faith in me. "Remember, darling, I'm counting on you." What if she could see me now—a miserable, gutless weakling!

I pounded my clenched fists on the gunnel. I would not let her down! Rage exploded inside me, and I found myself standing bitter and defiant, screaming curses at the sea, fighting it like an invisible foe. As anger began to burn away the fears that had paralyzed me I sank down and clutched the seat. Where there was a will, there must be a way. If only I could get Bertha out of the fiendish trough and turn her bow into the wind, she might stand a chance. Perhaps then, after lightening her load, I could crank the motor. But how? I wondered. How?

The next minutes blurred. As I probed frantically for an answer, I swept up the bucket and began to bail with mechanical fury. I tried to recall epic voyages mariners had made single-handed, the way their tiny vessels rode out violent storms. Like a flash it hit me—a sea anchor! The bucket? Yes, that would surely drag her bow upwind. But what would keep it from sinking? My eyes combed the turmoil in the bilge. The extra life jacket—hurry!

Heart thudding wildly, I crawled forward dragging bucket and life jacket behind me. Cursing my lifeless fingers, I tied the works to the bowline and flung it over the side; held onto the bow seat and waited, afraid to look. Bertha quivered, froze, and finally came about with a jolt, her innermost parts groaning. My hands were wrenched from the seat, my body crushed up against the sharp-stepped planks of the hull.

For a moment I lay limp and helpless, the breath knocked out of me, while the wind shrieked like a demented spirit. Then I realized I could no longer feel Bertha pitching or hear her planks moaning. When I looked up, the rails were bare and glistening, free of the foaming water that had surged over the gunnels.

Struggling to my feet, I wiped the hair from my eyes and watched the bow rise and fall, smashing the wave-tops and flinging out arcs of spray. Thank God! Bertha was riding the rollers like a gallant overweight lady.

Weak from strain and relief, I began to worry about where we were. The blurred figures of my watch showed four o'clock. Darkness would come soon.

Once again I pumped furiously. As the bitter cold slowly drained from my body, I searched the storm-tossed sea for land, a ship, even a bird. There was nothing—only the damnable, howling gray void. Minutes passed while my hands went on working like a piston. I stared at the motor, hoping to God it would still go.

Finally, though the floorboards were still floating, I could restrain my curiosity no longer. I gave the motor a hefty pull. Silence. I yanked again. Silence. Exasperated, I checked the controls: throttle forward, choke down; everything seemed perfect. Again and again I pulled. Not even a cough. I was scared, too sick and confused to think what to do.

Perhaps the ignition was wet. When I ran my hands over my damp clothes, I felt the warm lumps of the life jacket. Savagely I dug my nails into the dry Kapok and began rubbing it frantically over the wires, plug, and magneto. Now I heaved on the rope again and again. But the motor still ignored me. I dropped onto the seat half-crazed, exhausted, my hands bleeding and the smell of sweaty rope sickening me.

Then, as if the fates were not satisfied, they lashed out again. From downwind, the roar of breaking surf came to my ears. A hundred yards away, the sea boiled in a roaring turmoil of broken white water, exposing huge clusters of shaggy brown rocks like the hoary bodies of thrashing sea monsters. My breath froze in my throat. There was no sign of land, no way of telling where I was or which direction the wind and tide had taken me. All I could see, hear, and feel was the thunder of surf creeping toward the boat, filling the air with wet, choking mist.

Hovering over the motor, I cranked and cranked, struggling for balance under the wild rolling and lurching, afraid the last few strands of rope might break. Still not even a cough. Something was awfully wrong, but my blurred mind couldn't figure it out. The thunder of the seas boomed louder. The reefs were barely three boat lengths away!

In despair I sank beside the motor again, mumbling, cursing the sea. Mechanically I began to wrap the frayed rope around the flywheel. For some reason I noticed the direction I was winding it. Then I remembered Mr. Goodwin's warning. "You fool," I cried aloud, "you've been cranking the wrong way! "

With a single heave in the right direction, the motor started and Bertha shot forward full throttle into the wind, plowing into the chaotic waves. I clawed my way to the bow and pulled in the sea anchor, almost unable to believe my eyes. Rocks were so close I could nearly touch them with an oar. Rushing to the tiller, I held the bow into the wind and kept it there, watching with infinite relief as the acres of wild white water gradually fell astern.

Daylight was running out on me. I hadn't a clue where I was or if I had enough gas to reach Fernwood. But the sea gave me little chance to think about my troubles. Ahead of me, charging out of the grizzly gray murk, towering rollers bore down upon us. The wind rose still higher, smothering the sound of the hissing exhaust.

Bertha seemed to sense that her captain was in command. She leaped over the frothy crests and hurtled down the slopes into the troughs. Each time I thought she'd never raise her head. But each time she did, quivering and groaning, scooping in more sea over the bow. Then with incredible nonchalance she slogged upward, exhaust booming, girding herself for the next proud leap into space and wild downward ride as though she loved every second of it.

For a half-hour, I headed into the unknown while I continued to lighten my boat, thankful that the wind was dying and that I was still alive. Patches of blue streaked across the somber sky. I began to breathe easier as the waves stretched out in gentle ruffles that glistened in the westerly breeze.

Three miles away, lumps of land rose from the sea, wild and strange and brilliantly green like fragments of forest that had broken away from some far-off shore. My eyes feasted upon them. These, I thought, must be the Princess Islands. What I had suspected all along must have happened: the storm had blown Bertha northward, in their direction. But I soon discovered the gale had played tricks. Not only had it whipped Bertha to the north; it had lured her to the east as well.

I wondered if one of these islands was Salt Spring. By now, I knew, Jeanne would be at the end of the wharf there, anxiously waiting for Bertha to come down the Trincomali Channel. Undoubtedly she too had felt the storm on the ferry to Salt Spring.

I looked for the chart. It had met a watery death; all that remained was bits of paper sticking like confetti to the floorboards. Now, with land in sight, there was only one thing to do. I pointed Bertha toward the largest island in the hope of finding a settlement there. I tried to kid myself that the situation was not as bad as it seemed. The sea had ceased to be my enemy. And since the days were longer in the north, there were still several hours of daylight. Most gratifying of all was the glorious feeling of confidence that raced through my veins. Bertha was truly my ship now. In my own blundering way I had won my stripes.

Before long, the island mushroomed to starboard, and gulls wheeled overhead to greet us. I scoured the virgin shoreline for signs of habitation. Hunger pains were gnawing now. A hasty search had proved that nothing remained of my little brown sack of lunch except an apple, which I devoured in a half-dozen bites.

Then I spotted a boat that had rounded the nearby point. It lay half a mile offshore, almost dead ahead, as though it was anchored there. I gave Bertha a burst of speed.

It was a tugboat pulling two booms of logs, the wash from its propeller churning the sea white at its stern. I edged Bertha along the lee side, coming as close to the pilothouse as possible. A tall blue-jacketed man with a visored cap stepped out on deck and looked me over curiously.

"Where are we?" I yelled, cupping my hands to make my voice heard over the drumming of the tug's engine.

The man disappeared into the pilothouse and returned with a megaphone. "Turn Point, Stuart Island," he shouted through it.

This meant nothing to me—it was not even vaguely familiar.

"Where's that?"

"The San Juan Islands, U.S.A."

The gale had blown Bertha eastward across Haro Strait, over the invisible Canadian border and right into the lap of the American San Juans. It was almost sundown. I hated to think how far off Salt Spring—let alone Fernwood—might be.

The man on the tug apparently noticed my forlorn look.

"You'd better come aboard," his voice boomed. "You look like you're lost."

Before long I was telling my pitiful story to the skipper over one of the tastiest cups of coffee ever brewed.

He beckoned me to the window. "There's Salt Spring—that biggest hump across the strait, behind the little ones. You're not so far away."

My heart leaped in relief.

"Head for Beaver Point," he went on, lowering his finger. "Behind Moresby, there. It's about eight miles. Then you'll see the Trincomali through Captain's Passage."

For an instant I gazed hungrily at the distant blue hills of Salt Spring. Beyond them—somewhere—were Fernwood and Jeanne. It was past dinnertime already, and I had never stood her up before.

With a full can of gas, I thanked the skipper and climbed into the lifeboat, well briefed and pleasantly stoked for the final leg of my voyage. This time I remembered to crank the motor clockwise. Bertha plowed ahead, full throttle, her bow pointing toward Salt Spring as though she knew she was racing against the quickening night. The islands crept closer, each taking on a definite hue and shape of its own. I was glad the wind had swept the sky clear. During the next hour or so, I would need all the twilight the heavens could spare.

A half-hour later, as I neared Pelorous Point on Moresby Island, the sun faded behind a silhouette of ghostly mounds. Dark patches of forest began to close around me. Like tarnished emeralds, island upon island reeled into view, etching the soft blue horizon with their black feathery crowns. Wooded shores merged and black-marbled channels sprang open, only to close as new corridors appeared. The air held a tantalizing scent of woods and sea. Over the stern, pink-caped Mount Baker gazed down serenely, monarch of all that lay at her feet. Here, I thought, was a perfect marriage—a happy union of land and sea.

By eight o'clock, Bertha had skirted Moresby Island, and the shadowy tip of Beaver Point lay three miles ahead. Fernwood still seemed a long way off. Time was outracing us. My muscles tightened as I huddled closer to the tiller in the dwindling light. The massive black hulk of Salt Spring looked like a sleeping lion across the bow. I was afraid to shift my eyes lest it steal off into the night.

Dusk came swiftly, and with it a sickening chill of uncertainty. How could I creep through these islands in the dark? Even in the daytime, it would be difficult enough. And here I was, trying it at night without even a chart, light, or compass to guide me.

Moments later darkness swooped around me, sealing off the last glimpse of Salt Spring, the islands, everything. I groped fearfully for the motor, idling it as slow as I could, hoping it wouldn't stall. Standing now, peering into the blackness, my hand choking the tiller as I listened to Bertha's weird booming, I was overwhelmed with a sense of danger. I grew panicky—hemmed in by night and the invisible shore. I felt bitter and lonely—lonelier than I had ever felt before. All my childhood fears of the dark came rushing back. Whatever courage I'd had began ebbing away in the blackness.

Only a fool, I thought, would attempt to go on. It might be wiser to pull in to shore and wait for morning, rather than risk these rock-strewn waterways in darkness. Then I remembered the tide—I'd never get the boat off at daybreak. So I crept ahead blindly.

Minutes trickled by while Bertha's booming went on shattering the night. I could almost feel the damp scent of evergreens. I prayed for a light to blink out from shore to guide me, but beyond the eerie red glow of the exhaust, there was nothing but a black void.

I had to guess my position by the sound of the exhaust bouncing back from Salt Spring. When the echo took a fraction of a second longer, I knew I was farther from land. This was not good, for I remembered that on the chart the Channel Islands, scattered fists of rock, lay a half-mile off Salt Spring, somewhere to the starboard. I didn't dare swing too far out for fear of hitting them. To go on, I had to feel my way by ear.

Then I found that by squinting my eyes I could make out the towering bulk of blackness to port, a shade darker than the sky. All around me, now, the ebony sea sparkled with diamonds. Eerie shapes loomed up, dissolving into the dark as soon as I swung the tiller. Before long, I realized I was hopelessly off course and in a quandary. My ears had lost track of the right echo. I tried zigzagging, flinching and dodging the shifting shadows that leaped across the bow. Any moment I expected something dreadful to happen.

My watch glowed eleven-fifteen. Poor Jeanne must be frantic. In my mind's eye I could see her pacing the flimsy wharf, her flashlight stabbing the darkness and her ears probing the night for the sound of a boat.

In the next instant, I thought I had lost my senses. A ghostly apparition sprang up along the floorboards. Its shadowy form crept out in front of me, lengthening and falling across the motor, tapering off toward the bow. It must be a trick of my frayed nerves, I thought, and spun around. There, climbing like a silver sun above the sea's blackness, was the brightest moon I could remember. A sense of humbleness welled up inside me. It was as if God had sent a helping hand. Hope soared, warming my chilled body and aching limbs. This was no longer the forbidding land I had battled all day. The moonlight painted everything with silver splendor. Hills of white forests now circled the boat, and the air exhilarated me with its tang of low tide and seaweed.

I was farther from Salt Spring than I had thought. The Channel Islands squatted over the stern hardly a breath away. I gazed longingly to the east where I imagined Captain's Pass to be. There in the moonlight between Salt Spring and Prevost, the narrow gap glistened: two white fingers of land stretching into the sea. I swung Bertha toward them with a burst of speed. Off to port, the lights of Ganges twinkled in the night like distant stars.

Moments later, I throttled Bertha to a crawl, barely able to contain my excitement, wondering if Fernwood wharf would be visible on the other side. Every second seemed like a minute. Behind me, I could hear the wash from Bertha's wake pounding the rocky shore as we crept through the pass and rounded the shadowy point of Salt Spring into the Trincomali Channel. Far off in the moonlight, a low dark arm projected into the sea. On its tip, a faint light flashed off and on.

CHAPTER 4

Bridgehead on Eden

IT WAS ON HER MAJESTY, QUEEN VICTORIA'S BIRTHDAY—the twenty-fourth of May—that the trek to our new island home came to an end. For weeks we had lived for this day. Now, from Fernwood, we were eager to strike across the channel and forge a bridgehead on our dream.

The morning was bright and calm. When we started to leave the Bettiss farm, Vic shoved something into my hand. "Take this tidebook, son. Better make it yer Bible—got an idea you'll be needin' it."

As we walked toward the car, loaded with eggs, butter, milk, and a jug of water, Vic mentioned casually that he had been out early, inspecting Bertha. She was grand for our purpose, he declared, except for one thing. "That's a lotta boat for that pidley engine. Watch it, Tiny. She may git you in trouble."

I had reason to recall his words later on.

We were lucky to know Vic and Myrt. Their hospitality was so warm and genuine that we felt as if we'd been friends all our lives. We leaned on them as guides, too. Pioneers themselves, they had grubbed their land from forest near the wharf and settled back to farm. They were a simple, kindhearted couple, content with their world, such as you might find in any remote area. Their weathered home, built from the timbers of a discarded schoolhouse, hugged the soil with a staunchness that reflected their character.

We waved good-bye to the Bettisses and backed down to the wharf to unload the trailer onto a small creaky cart. The iodine scent of low tide filled the air. Stretching far out under the pilings, the mudflats heaved and glistened in the sun. But the sight of the island quickened our hearts and soon made me forget yesterday's ordeal.

We shucked off our sweaters, and wheeled the first load to the end of the ancient four-hundred-foot pier. From there, everything had to be lugged down the steep gangway. We shuddered at the perpendicular drop of nearly thirty-five feet. The desk we slowly slid down the mossy gangway railings, but I rolled the Singer down on its wheels, holding onto its legs. Neither of us dared get in front of it.

"Don't let go, for heaven's sake," Jeanne commanded, a bit skeptical of the operation.

I clutched the railing with my right hand and inched the machine downward a little at a time, bumping it over the worn cleats of the ramp. Suddenly a chunk of railing came off in my hand.

"It's getting away," I heard Jeanne yell. "Grab it—hurry!"

Frantically I made a dive for the machine and grasped its cover. We went crashing and bouncing down the gangway together, banging into the piling that stood in the center of the float. I felt my head ram against the steel-latticed legs. The next thing I knew, Jeanne was bending over me and wiping my face with a wet handkerchief.

"Speak to me, Dave. Are you okay?"

Groggily, I rubbed the bump the size of a golf ball that was splitting my head.

Jeanne dropped my arm and rushed over to the Singer. "I'm glad that piling was there. We'd have lost her for sure."

"Been better if we had," I snorted.

After the fifth cartload Bertha was bulging, but the trailer was still half full. By then, we were no longer alone. It was a holiday, and the natives in the vicinity, out for a stroll or to fish, had dribbled onto the wharf. Or they had heard that Americans were moving to Wallace Island. Several old men in baggy pants and threadbare coats stood on the pier sucking their pipes, watching us load with solemn curiosity.

The audience made us uncomfortable. And I felt even more so when I looked at poor Bertha, with our desk, dresser, and chairs piled high across her at midship. Were we overloading her, I wondered? I didn't want anything to spoil our maiden voyage.

When I picked up the last box, one of the old men stepped forward with a puzzled look. "All my life I've dreamed of living in California," he said. "Here you are—giving it up to live here." He was plainly bewildered.

Jeanne, arms loaded, elbowed my ribs as I stood blocking the gangway. "Come on. We'll never get everything over before dark."

While we filled every cranny in the boat, more natives sauntered onto the pier and stared, talking among themselves. Odd bits of conversation floated down to us.

"Say, chappie," a man in a tweed coat called, "where's your yacht?"

When I replied that we didn't have one, a murmur of disbelief came from the spectators. "An island without one of those fancy Chris Crafts?" we heard someone say, and another man declared, "They're probably just the owner's kids—or caretakers, maybe."

"They sure think we're crazy," Jeanne whispered as she settled a load of groceries on the transom seat. It was almost noon when, ready to go, we discovered we had forgotten to leave room to crank the motor. I dug out the box of cooking utensils and squeezed it onto the seat against the steering cable. Jeanne took the tiller as I shoved Bertha off and started the motor. Bertha roared and leaped forward, bringing the pile of furniture down onto the motor, smashing the throttle full forward. Turning to clear the debris, I heard Jeanne yell, "The rudder is jammed!" My head swiveled back toward the bow—we were heading straight for the mudbank!

Jeanne frantically shook the tiller, but it was too late. Bertha ground to a halt in the mud. With faces flaming like overripe tomatoes, we kicked off our shoes and slipped over the side. Howls of laughter came from the pier as we worked to push Bertha free.

Finally, we headed toward Wallace. Jeanne straddled the bow, her blue denims rolled to the knees and her legs splashing the sea. Our eyes began searching for the little beach near the shack as soon as the cove opened before us. Tingling with excitement, I cut the motor, and we glided aground in the shallow water.

I swept Jeanne off the bow, her face beaming like a bride's, and waded ashore with her. "Your kingdom, darling," I announced, planting her feet on the soft white shell.

We hugged and kissed each other in joy, then just stood and gazed about rapturously while we drew in huge lungfuls of unbreathed air spiced with the aroma of sun-drenched cedars. The silence was so complete we could hear our hearts beat. We felt like royalty surveying a new empire. Wallace was our world now—this cove with its tiny beach cradled between upthrusts of sepia sandstone and its great firs melting into the sky, sheltering its secret waters.

"Dave, does anyone have the right to own anything this beautiful?" Jeanne asked seriously.

"Yes, if he earns it," I answered with equal seriousness. And I began to think of the hard work ahead, of the things that had to be learned and done in so short a time, to put the island on its feet. Our determination and toil were the real price of the island.

We pulled Bertha farther up and put down planks of driftwood so that we could scoot and tug the heaviest furniture and boxes onto the beach. Unmindful of our presence, ducks skittered onto the cove and crows and gulls fought for starfish among the barnacled rocks. The sun was so hot it made our necks and arms sting. By the time Bertha was emptied, we had stripped off most of our clothes and were secretly itching for a swim. But it was nearly three o'clock, and nothing would stop Jeanne from getting everything moved to the island before dark.

By sundown, with all our possessions ferried over, we stood on the beach surrounded by countless dark pyramids, almost too tired to find something to eat. A bluff sprinkled with madronas and salal overlooked the spot.

Jeanne pointed to a bare area of shell. "Let's camp here. We can lug the stuff up in the morning."

"You start supper," I said. "I'll rustle wood for the fire."

It was a glorious night. In the confusion we couldn't find most of the groceries, the camp stove, or the flashlight. But that didn't matter. We heated a can of spaghetti in the coals, munched soda crackers in place of bread, and washed it all down with coke. We were so happy that it tasted better than any feast at Perino's. After we finished our cigarettes, we zipped the two sleeping bags together and crawled in before the fire to watch the flames dance and crackle in the darkness. Burning driftwood filled the night with an oriental scent. A few feet away Bertha rocked gently at anchor, and overhead the vast sky gleamed with stars.

Neither of us could remember having felt our existence so worthwhile. Our neighbors on Salt Spring, we knew, were skeptical of our undertaking, especially since we had come from California, the golden land of their dreams. Strangers are always closely scrutinized by islanders, for gossip is their chief amusement. We had furnished them with the choicest tidbit in years.

The magic of the night loosed all our thoughts. We planned and dreamed of the island for the days ahead.

When a distant loon gave a mournful cry, Jeanne nudged closer. "This doesn't seem real. I've never been this happy before—have you?"

"Never." I put my arm around her. "You know, there's something sad about happiness. It's like a breath of air—you can't hold onto it long."

Jeanne rose to her elbows and turned toward the fire. Her voice became wistful. "Somehow, I don't feel like the pioneer type. Dirt has always bothered me.... And there's something else too—a man is always seeking and building. He really never has what he wants. But a woman usually does. Her job is to keep what she's gained." She studied my face. "You have such great plans, Dave. I don't ever want to be a burden to you."

"What foolish nonsense! What would I want with an island without you?" And I kissed her gently.

We lay back, close as two people can get, and looked up at the stars. For a moment, the warmth of our bodies seemed to dispel the hardness of the cold ground. A light breeze played against our faces, while a far-off plane droned across the pale night sky.

But sleep eluded us. The ground soon became a sheet of iron. We tugged and twisted at the sleeping bag to keep the cold from our shoulders. And as the fire died into glowing coals, the trees lost their friendliness and began hovering about us like sinister figures. An owl hooted nearby. Every sound made the silence that followed more frightening. Then a horrible croaking noise shattered the quiet.

Jeanne threw her arm around me. "What's that?"

The faint outline of a huge blue heron sailed past, flapping ponderous wings like some ungainly prehistoric bird.

"Only a heron."

"I'd feel better if you'd put more wood on the fire," Jeanne said. "It's getting so black."

I threw a few sticks on the coals and we settled back once more. A moment later, Jeanne raised her head to listen. "What is it?" I whispered.

"Drums! Don't you hear them?"

I listened too. A weird pulsating rhythm throbbed like a giant heartbeat. The sound seemed to permeate the darkness.

"There's an Indian Reserve not far away." I hated to tell her, remembering the carpenter's story, but wanting to soothe her fears and my own. "Nothing to worry about. They're just celebrating the Queen's birthday, darling."

In a few minutes the drums stopped and an eerie silence fell over the cove. At last the tension began to drain from our bodies, and as drowsiness enveloped us we curled up in each other's arms. Dozing off, all I could mumble was "Uh-huh, uh-huh" to Jeanne's final question: "Do you think Bertha is out far enough?" Like a true woman, she wanted to protect her earthly possessions even while she slept.

During the night I was awakened by the pattering of little feet across the foot of the sleeping bag. Off and on, beady eyes glared like beacons in the blackness. Pack rats were twittering among themselves: Where did this wonderful food come from, and who are these strange invaders?

The next thing I knew, Jeanne was shaking me awake. "Look—Bertha's high and dry!"

I struggled to my feet, squinting in the bright sunlight at the nearly empty cove. Poor Bertha squatted in the mud like a crippled bird. What a stupid blunder I'd made, forgetting to check the tidebook. If Vic could only see us now.

"Well, don't stand there," Jeanne scolded. "Can't you do something?"

Guiltily, I inspected our floundered craft, surrounded by noisy onlookers—crows that swept low overhead, cawing raucously, mocking and laughing, darting from one fir limb to another. I shook my fist at them, but it did little good.

"She's okay," I shouted above the ruckus. "All we can do is wait for the tide."

The cove had shrunk to a quarter of its normal size. The two sloping arms of the entrance stood barely forty feet apart, like a gap in a crumbling stone wall. Only drift marked the high-water line. I dug out the tidebook. By the figures, the cove had dropped thirteen feet since the evening before. Where did all that water go? Now I realized what a job it was going to be to become independent of the tides. We would have to construct a little wharf with a ramp extending out far enough so that at low water the float would never go dry.

The air tingled with the aroma of drying kelp, and in the mud flats tiny geysers of water shot up between upturned shelves of sandstone.

"What are those spouts?" Jeanne asked.

"Clams."

"Wonderful! I've never tasted real chowder."

I pointed down the beach toward Bertha. "See those knobby white bumps on the rocks? Oysters. Scads of them."

After a breakfast of oranges, bread toasted over hot coals, and instant coffee, Jeanne discovered the telltale signs of thieves during the night. We followed a trail of flour to a grocery box. Hardly a spoonful was left in the five-pound sack.

We began at once to break the hundred feet of trail up to the shack. "Here's your palace," I said, shoving open the door, which groaned and promptly crashed to the floor in a cloud of dust. Rats the size of fat cucumbers scampered about for a place to hide.

Jeanne clutched my arm as we stepped inside. "It smells even worse than it did before."

With a knife, I pried out the two tiny windows and laid them on the floor. The sudden splash of light seemed to magnify the chaos. Cobwebs sagged from the weight of fresh dust, and rat droppings anointed the table and floor. We stood over the ancient range, the word "Majestic" still visible on the oven door.

Jeanne started to touch it, but withdrew her hand. "I don't know how I'll ever cook on this monstrosity."

"Old Bessie can wait," I said. "Let's put up the tent." Jeanne's face brightened.

Above the beach we cleared a spot under the madronas, then combed the tideline for planks and boards to put down as a floor. By noon, we stood inside admiring our job. For the next hour or so, we slipped and stumbled as we toted up the trunks and furniture, which seemed to have picked up pounds overnight. The hope chest was too heavy to lift, so we took out the linens and bedding and packed them up separately.

Once the most valuable things were stored in the tent, Jeanne dug into the food boxes. She came up with peanut butter sandwiches and tomato juice straight from the can. We were just swallowing the last drop when she sprang to her feet.

"The Singer! We nearly forgot it."

I grinned. "Maybe the tide took her out last night."

No such luck. She was still buried under a mound of canvas. Remembering how the Filipinos on Leyte carried their machines from house to house, I cut a sturdy fir pole and, with a bit of rope, strapped the Singer to the center of it. We hoisted the pole to our shoulders and marched up the bank to the tent like native bearers.

By five o'clock the beach was empty and the tide nibbled at Bertha's keel. With Jeanne's clothesline and pulley, I rigged a mooring buoy using a chunk of cedar driftwood. Attaching the line to Bertha's nose, I pulled her out to the buoy.

"There's more Crusoe in you than I thought," Jeanne said when she saw my accomplishment.

Pork and beans had never tasted so good as they did that night. We unrolled our sleeping bags amid the boxes and crates in the tent and crawled in, totally exhausted, in the glowing twilight. No drums or eerie noises could have kept us awake—we slept soundly until the clamor of crows routed us out at daybreak.

After breakfast, we decided to tackle the shack. We hauled up buckets of seawater to scour the floor and dumped piles of rubbish on the beach. With the dirt and the shreds of brittle tarpaper gone, chinks of light seeped through the rough shiplapped walls. Jeanne declared war on the cobwebs, and dragged down bird nests from the rafters, disturbing a colony of bats, which sent her running out in terror.

By late afternoon, the worst was over. The shack looked hospitable, the beachwood door was hung back on leather hinges cut from suitcase straps, and after a rubdown of sandpaper Bessie wore a respectable smile.

"Doesn't look so bad now, does it?" Jeanne said as we surveyed the tiny boxlike room. "Think we can move in tomorrow?"

"If we work tonight. There's that rotten place in the floor to patch, the cupboards to hang, the windows to—"

"And a counter to fix," she interrupted. Like a couple ready to move into a plush new ranch house, we could hardly wait.

While Jeanne wrestled supper, I pried open a heavy wooden crate, almost tasting the smell of new oily metal as carpentry, masonry, and plumbing tools glistened before my eyes. I felt like a child with a boxful of new toys. Not a one had I used before. From now on, books had to be my guide and experience my teacher. I examined a plane, a pipe cutter, wrenches, and a trowel, wondering how quickly I could learn all these skills.

After supper, we hung a gas lantern from a rafter and attacked the cabin again. By nine, working doggedly, we had patched the floor, slapped together a makeshift counter of beach planks, and put back the two windows minus their cracked panes. Then we brought in the unpainted kitchen cabinets and screwed them fast to the walls above the counter. The shack was to be the kitchen-nucleus for our house. With a tape and yardstick, we went on planning and measuring until shortly after midnight, when we finally gave up and retreated to the tent. Long before our bodies had warmed the cold damp sleeping bags, we were sound asleep.

Morning came bright and clear. We emerged from our warm cocoon to the familiar rhapsodies of the crows. We found ourselves enjoying the antics of these noisy friends. At breakfast, they wandered about our feet and squabbled over breadcrumbs in the grass. We named one Freddie. With a staccato voice like Spanish castanets, he serenaded us from his perch in a madrona limb.

Moving into the shack was the first order of the day. Before long, we had a well-worn trail to the door. The walls soon seemed to be shrinking, for each bit of furniture became more difficult to squeeze inside than the one before. The floor groaned as I shoved articles through the door and Jeanne stacked them. It was as if we had dumped the contents of a five-room house into a one-car garage.

When Jeanne complained that there just wasn't room for all the stuff. I assured her that we would manage somehow. "We'll have a living room, bath, and bedroom on this place in no time," I promised bravely. It was just as well I could not read the future.

That night we celebrated our first meal in the shack. Jeanne opened a canned chicken and heated it on the camp stove, which was perched on the range. We attacked the steaming platefuls from the top of the nearest crate. The room was still warm from the afternoon sun, but the evening sky was barely visible through the checkered windows, which looked like ice-cube trays planted in the walls. We were disappointed they were so high we couldn't see out while we ate. But in spite of the frustrations and the confusion we were happy, and it showed, along with the chicken, on our faces.

The food relaxed us. Right then, the thought of any more work was distasteful, so we swung open the door and sat on the worn threshold, smoking and watching the purple twilight filter through the fine lacework of the dark firs. How still! Sound and time had ceased to exist. The clamor of the life we knew was gone. We had turned our backs on civilization, hoping that—someday—the world would come to us.

"Tomorrow," said Jeanne, "we'd better clear out the view." She nodded toward the cove. "I'll feel easier when we can keep an eye on Bertha."

CHAPTER 5

Funny Wooden Blocks

THE LANTERN HISSED DOWN A WARM GLOW FROM THE RAFTERS. After the dishes were done, we assembled the bed near the door. How good an innerspring mattress would feel tonight! While Jeanne filled the cupboards with dishes and food, I studied the tidebook and charts of the area, checking the depths and currents around the island. There was a deep passage through the kelp beds at the entrance, but what surprised me was the speed of the Trincomali Channel. At peak tides, it raced by our doorstep at a five-knot clip—nearly as fast as Bertha went. That worried me. What we suspicioned now secured a fact: To an islander, the greatest of all hazards are at sea.

For the lonely islander leads two lives—he belongs to both the sea and the land. The word "islander" is far too nebulous. He is a sea-lander. The salt chuck is his highway and lifeline, lumberyard and wastebasket, his means of livelihood and supper, fertilizer and fuel, the fence around his yard, and—often—his cemetery. Like the isolated coast dweller along the mainland, where the mountains plunge into the sea as carelessly as gulls dot the sky, carving fjords, inlets, and sounds, he depends for his very existence on a strong sea boat. The sea-lander diligently protects it, fondly honors it, and faithfully nurses it, for the stakes are high. A wrong decision anywhere else can lead to ruin; at sea, the answer is death.

The next morning we missed being awakened by the crows.

"It's late," Jeanne said, rolling out of bed. "Funny how the crows disappear all day. Only a few like Freddie seem to stick around. Where do they go?"

"Foraging to the other islands, I imagine. They're commuters, you know." The suntans felt cold as I put them on. "Smart, aren't they? With the pick of the islands, they choose this one for home."

"Aren't they generally considered a bad bird?"

"No bird is really bad—only the people who kill them." I leaned over to tie my shoelaces. "I couldn't shoot one—or any animal, for that matter. Even as a boy guns didn't appeal to me, and the war didn't help—"

"Come here—quick!" Jeanne motioned from the window.

There, grazing under an apple tree, was a doe. The color of her velvety coat matched the dry brown grass that she nibbled. Barely breathing, we watched her, entranced. Then my elbow struck something on the cluttered shelf. The doe's head sprang up, her ears alerted, nostrils sniffing the air. The next moment, she bounded silently into the bush.

"Right here on our island. I can't believe it.... How could anyone harm such a lovely creature?"

Jeanne had read my thoughts. To me, hunting with my camera was far more exciting and civilized. Then and there, we proclaimed the island a sanctuary where no living creature need fear any man, though neither of us knew how difficult this pledge would be to keep.

After breakfast, Jeanne held up the water jug and shook it. "Empty," she said.

"Come on—we'll look for the well. Something tells me it's not going to be easy to find."

We started the search with a circular sweep around the shack, hacking and probing the bush as though we were looking for signs of buried treasure. And, in a way, we were. Water is worth more than gold to a sea-lander. Next to his boat, it's his first concern. The lack of it, we sadly discovered in the days ahead, could break our backs and hearts and ruin our dream.

By eleven o'clock our hopes began to dwindle. Hot and weary, we were no closer to finding the well than when we started.

"This is pretty rough going," Jeanne declared as she came up to me, wiping her face with her arm. "Why don't we wait till it's cooler?"

"We better keep on," I muttered stubbornly, the sweat trickling down my back.

Wherever we wandered, in shade or sun, salal jacketed the ground, covering rocks, gullies, clearings, and ridges with a thick green blanket. It hugged and choked the largest fir trunks as much as five to six feet above the ground. Only a few patches of rock broke the dark greenery. Under its waxy leaves the prolific canes grappled with one another, intermeshing into wicker cages, unaware of the treacherous devil's foot coiling upward to strangle them. Every step held danger—the possibility of a twisted ankle or bruised leg. As a precaution, the salal had to be thrashed back to show the ground. Even then, you couldn't be too sure.

Noon came and went. Fanning out, we scouted several hundred yards from the shack and found more apple trees swallowed by the bush. My jaws clamped tighter as I drove on blindly.

"Don't be in such a rush," Jeanne called. "Let's rest a minute."

But I kept plunging ahead. All at once the earth crumbled under my heavy step and I felt myself falling, a whirl of stale air engulfing me. My arms flailed out wildly for something to grab. Then I hit water.

Gasping for air, I fought to the surface and instinctively began to tread water. In the moldy darkness, my hands struck against slimy clay walls. I clawed frantically at them, but my nails could not dig in. Far above me was what looked like a pinhole of light. It vanished immediately as Jeanne's voice rang out.

"You all right, Dave?"

The darkness reeked of decay. Gripped by the intense cold and nausea from the awful smell, I struggled to keep my head above water.

"I think so . . . but I can't keep treading water much longer."

"What'll I do?"

"Get the life jacket and rope . . . from the boat," I sputtered, pumping legs and arms with all my strength. "Hurry!" I could see the outline of her face now, hear her gasping breath.

"Wait—I'll be right back." Her voice faded with the sound of her footsteps, and a finger of daylight again punctured the clammy blackness.

Wait! That was a laugh. What else could I do?

My breath caught in my lungs, and it began to hurt to draw in air. I knew I was tiring fast. The water climbed higher up my neck. To save my legs I paddled harder, whipping the water into foam, but my arms soon grew too heavy. When I lashed out at the mucky walls, I sank—and had to surface and battle again for breath.

The minutes blurred. Panicky and exhausted, I wondered how much longer my legs would last. Where was Jeanne? I was afraid of cramps, afraid of—but I couldn't think of that. There was too much to do, to drown. What would become of Jeanne? Our island dream? I must keep moving.

More minutes dragged by. The water crept hungrily over my straining chin... into my mouth. Choking, I spit it out. Oh God, I couldn't go on.

All at once the light was blotted out.

"Here's the life jacket, Dave!"

It splashed by my side, and I pounced on it, rolling it under my body. The jacket felt warm against my icy skin. What a glorious relief to stop bicycling and relax my arms! Paralyzed with cold, they lay draped lifelessly over the jacket.

"You okay?" Jeanne, panting heavily, poked her head through the hole. "You're so quiet."

Stones and bits of loose dirt plunked around me.

"Yes, okay now." I heard her deep sigh.

After a moment I struggled into the life jacket, forcing my wooden fingers to buckle it across my chest. Then the thought struck me: How could she get me out of this stinking dungeon before I froze to death? I gazed up fearfully, gritting my teeth to stop their chattering. Jeanne couldn't possibly pull up two hundred and twenty pounds for what must be a good twenty-six feet.

She echoed my thought. "How can I get you out, Dave?" And the fear in her voice matched my own.

"Toss me the rope—I'll try it." I had forgotten how blistered my hands were and that I was considerably overweight.

I lunged up at the rope. Clasping it between my legs and straining every muscle, I tried to make a second grasp. The water gripped my waist like glue. I felt the rope burn through my fingers as I sank back in the water.

Jeanne leaned over the hole. "What's wrong?"

"Damn it, I can't climb this miserable thing!"

She groaned and disappeared. Suddenly the rope tightened in my hand, and I had to duck a shower of debris. "Hold on!" she yelled.

"Don't try it, honey—you can't do it." My heart sank. The poor kid!

"We've got to think of something else."

Her head reappeared, and we stared unseeing at each other. We could almost hear our minds working, trying to latch onto some scheme. Then I remembered the set of blocks I had brought for a block and tackle. I had planned to experiment in my spare time to figure out how they worked.

"Those funny wooden blocks with wheels in them," I shouted. "Can you find them?"

"I think so. What good'll they do?"

"Get 'em and I'll tell you."

She scrambled off.

Shaking with cold, I tried to picture how a block and tackle worked. Which way did the rope go in the pulleys? Where did you tie it? I'd just have to trust to luck that I'd blunder onto the right way.

Then I heard Jeanne's footsteps, and she was peering down at me again.

"Here they are. What'll I do with them?"

"Haul up the rope and thread it through the pulleys—from one block to another," I ordered, guessing. "Then tie one end of it to a block and pull on the other."

Moments later, she called, "All set. What do I do with the blocks?"

I hesitated, shivering so badly I could hardly think. It was almost impossible to concentrate and still keep arms and legs moving to ward off the cold. I felt as if I had been in this freezing hellhole for an eternity. My hands were like stiff claws, and I could no longer control the chattering of my teeth.

"Try hooking the block with the pulling rope to the nearest tree. Then lower the other one to me."

A rain of rocks and dirt splattered the water. When it stopped, I reached up at the weird assemblage of ropes and pulleys hanging over me and grabbed the block, thankful it had a big steel hook on the bottom.

"I've got it!"

I could hear Jeanne wrestle with the gear on top. Not trusting my thin leather belt, I tucked the hook through the buckles of the life jacket.

"Heave away," I shouted, covering my head with my arms.

Dirt showered into the well as the ropes tightened. The pulleys commenced to squeak, and the life jacket clamped about my shoulders and chest like a cold compress. The pressure was painful but good because it meant that the tackle was working and that I was going up at last. To make sure, I glanced down. My feet were dangling just above the water.

"How you doing?" I called, wondering if the strain would be too much for Jeanne's small body and slender arms. My eyes were squinting from the daylight. Now the hole was larger, perhaps fourteen or fifteen feet away. Still not a sound from Jeanne—just the scuffling of her feet and the squeal of the pulleys.

I dared not move a muscle. Only by the frightful pain in my shoulders as they grazed the stone walls did I know that I was actually making progress. More loose rocks and roots plunked into the water. Then my eyes fell on the straps of the life jacket. They were stretched thin and taut, on the verge of breaking. Quickly I reached up and grabbed the small block, strangling it with my hands to ease the strain on the straps.

"Dave! I can't—I can't go on!"

My heart leaped into my throat. The rope slackened in jerks, the blocks screeching, each jerk squeezing the air from my lungs.

"What shall I do?"

The top yawned like a splintered hole in a ship, seven or eight feet above my head. I grasped the first thought that came. "Is there a tree near you?"

"Yes."

"Work toward it-even if you lose ground. Then wrap the rope around it and rest." With forced cheerfulness, I added, "You're doing swell."

The pulleys squealed. Praying the life jacket would hold, I felt myself dropping in spasms, jerkily, as if I was being sucked back into the smelly blackness. Above me, I could hear Jeanne's groans and scuffing feet.

"I've got it," she cried, panting. "Hold on... I'll try again in a second."

Slowly, inch by inch, as if I was being hoisted by a wobbly crane, I came nearer and nearer the surface... five... four... three feet from the top now. Bright sunlight forced my eyes shut. Dirt, rocks, and chunks of rotten board rained down on me.

"How much more?" Her voice sounded panicky. "I can't reach... the tree."

"Keep going, honey—only three more feet. Don't stop now!"

Then it came. A heartrending rip in the life jacket. The straps were tearing away, peeling off long strips of nylon. My hands fought savagely to hold onto the block, but it slowly slipped from my grasp and the steel hook dug into my clenched fingers. I clung to it desperately.

"Pull," I screamed. "For God's sake—pull!"

My arms pained as if they were being yanked from their sockets. Fighting off faintness with each agonizing jerk, I knew in my blurred mind that I was inching upward, that I couldn't let go.

The next moment, I felt my wrists gripped in a vise. Jeanne had reached the tree, knotted the rope, and rushed to the hole. I stared up into her frantic, determined face. Then it dawned on me that I might pull her into the well. Swiftly, I freed my right hand and seized the ropes where they were partly buried in the earth.

Jeanne cried out sharply. Her face pinched in fear, she instantly grasped the other wrist and pulled on it with all the strength her small body possessed. I dug my toes into the earth to pry myself upward, at the same time clawing at the gritty ropes with my right hand. Groaning and pulling, like mountain climbers nearing the top of a craggy peak, we battled up over the crumbling edge and, in total exhaustion, collapsed on the ground.

Finally, Jeanne raised a pale, dirty face and started to cry.

"What's wrong? There's nothing to worry about now."

She dropped her head on my chest. "I know." Her tears rolled down unchecked. "I'm just happy you're safe. Every time I think of you down in that horrible hole, I want to cry...and crying makes me feel better." But she wiped her eyes resolutely and tried to get hold of herself.

I kissed her wet cheeks, wondering what I had ever done to deserve her, until she pushed me away and sat up with sudden determination. "Mr. Crusoe," she said severely, "I hope this teaches you not to go through the bush like a charging bull."

A crow chattered in amusement from a nearby fir, and I tried to hide my chagrin with a chuckle. Then, still weak and shivery, I shucked off my wet clothes and sprawled out in the sun.

Jeanne fumbled in her dirt-stained pants for two crushed cigarettes, and lit them. As she pressed one to my lips, I noticed her bruised hands. They were raw, badly skinned, yet she had not complained.

While the sun warmed my blood, I kept thinking about what might have happened if I hadn't brought the blocks. It seemed incredible that a couple of chunks of wood with wheels could be so vitally important. So this was what it was like to live on an island, to live or die by your own wits. Careful planning and meticulous attention to details were prime essentials, for apparently the easiest thing to meet up with on an island was trouble.

Before long, the sun slumped behind the firs. We sat up and examined our wounds, then scrambled back to the shack for the first-aid kit. Taking turns, we applied Band-Aids and bandages. When we were finished, Jeanne said, "Tonight we'll use paper plates. Neither of us is in shape to wash any dishes."

The next day, we piled into Bertha and began to explore the shore for sand and gravel. Fireplaces and foundations would take lots of both. Scouring the tideline, we guided the boat into every cove and inlet, marveling at the many eagles, herons, cormorants, and ducks that made the island their home. Wherever we saw the tiniest beach, we hopped ashore. When we chugged into the largest bay, a half-mile finger of sea driven into the forest, we momentarily forgot our quest in awe at its beauty. Jeanne exclaimed, "There's enough room here to anchor a Princess ship. Let's call it Princess Harbor."

By afternoon we had nearly circumnavigated the island and found nothing but small clamshell beaches. Discouraged, we skirted Panther Point opposite Fernwood, then pulled into a narrow inlet facing south, about three-quarters of a mile from the shack. Anchored to grassy slopes, giant firs and cedars leaned over the mirror-still water.

"There's a good beach." Jeanne pointed from the bow as we headed toward it. "Looks dark too, like gravel."

She was right. We named it Pebble Beach. Scooping up handfuls of the stuff, we let it spill through our fingers like precious stones. It never occurred to us to dig our toes down to see what lay beneath. All across the hundred-foot beach, gale-driven drift was piled up against a bluff of baby firs; the spot was almost inaccessible by land.

Nowhere, that day, did we find even a bucketful of sand.

"How will we make cement?" Jeanne asked on the way home.

"There's bound to be sand on Salt Spring. Maybe Vic knows where. Guess we'll have to pack it over."

That night Jeanne made the first entry in her new diary:

> MAY 29: The island is so exciting I hate coming in to fix meals. No rain yet, and the sun gets awfully hot with no smog to filter it. Lovely long twilights that last till ten. This morning, I thought someone was chopping our trees—just a redheaded woodpecker attacking a fir, but he was bigger than a parrot. Everything here is green—and big and beautiful. Dave and I can't join hands around the biggest firs.
>
> Still haven't got over yesterday's scare. Dave really butchered his hands. Today we went looking for sand—no luck. Means more hauling and less time for building—as if we didn't have enough to do already!
>
> Rats raided the flour and sugar again last night, and another trap is missing. They keep us awake, scuttling around in the walls. We've added a cat to the shopping list.

The first week of June found us comfortably settled. We had cleaned out the well, and at the bottom found one tan shirt, a Zippo lighter, four bits, and—luckily—the small leather pouch containing our two car keys.

From our west window, there was a perfect view of the cove and a garden patch started under the apple trees. Jeanne had transformed the shack into a home. She had curtained off a bedroom with our old Nile green drapes, laid scatter rugs on the floor, and hung a faded bedspread across a corner to create a closet. The desk served as a nightstand; the dresser stood against the north wall.

The ten-by-ten area that remained was our kitchen-living room. We could stand in the center and touch everything in it. Jeanne had brought in oyster shells for ashtrays and decorated the table with a centerpiece of driftwood and wildflowers. Pink ruffled valances hung at the windows. In fact, the little shack beamed with our belongings. And Jeanne beamed too, adjusting curtains and rugs as though she had never owned a house before.

The next morning, to conserve gas for the camp stove, we decided to let Bessie cook breakfast. I laid a fire in the firebox and dropped in a match. In seconds, we were choking from the smoke that billowed up from the plates.

"Hurry—open the windows," I ordered, grabbing the bucket and heading for the cove.

Whissssh! Bessie sizzled from the drenching I gave her. I retreated out the door, but after a moment we both crept back in.

"The flue must be blocked," I decided. "I'll take a look at the chimney."

By climbing the apple tree at the back, I managed to get on the roof. The chimney was a queer brick structure on wooden legs. I thumped on the roof and called to Jeanne to come up. In a moment she was beside me, peering into the chimney.

"Holy cow!" she exclaimed. "A bird's nest—and with two eggs in it."

The eggs were almost black, but we wiped them off and found they were really a pale blue. We pried out the nest carefully and set it in the crotch of the apple tree. But the mother never returned.

In a few minutes, Bessie was roaring beautifully and coffee was bubbling in the percolator. Not all our problems, however, were solved so easily. Our water was peculiarly hard.

"Sure lousy coffee," Jeanne said, wrinkling her nose. "Do you think it's the water?"

"Probably still brackish. It'll taste better after we use the well a while."

The old privy was another problem, and one we couldn't get used to. Its condition was such that life, limb, and dignity were at stake every time we forced open the door.

'Why don't you try out your new tools and build one?" Jeanne suggested. "I'll gather drift lumber. There's enough in the cove alone to do it."

I was inclined to agree with her. You can't expect a woman to empty a potty in the sea indefinitely.

Enthusiasm boiled inside me that morning as I grabbed the shovel and marched to the edge of the orchard. But digging the hole proved a far greater task than building the structure. After an hour or two, I began to recognize another lesson: Everything takes twice as long to do on an island, even diging a hole. Wherever I picked a pretty site among the madronas and willows, the earth was like hard rubber. I had to jump on the shovel, pogo-stick fashion, to drive it into the ground. I started half a dozen different holes only to hit bedrock in a couple of feet.

"Don't cover them up," Jeanne said. "I'll put the cans and garbage in them."

This was consoling. At least my effort wasn't totally wasted.

After three days of sweat, a bent saw, and a swollen thumb, the privy was ready for inauguration. We could sit comfortably in it without a quiver or a draft, though it was neither handsome, level, or square. But it had the most heavenly view in the world. The Courthouse, as we called it, was conducive to hours of deliberation on the wonders of nature. It stood under a canopy of slender, flesh-pink arbutus. We could watch the robins lunch off their red berries and see the cove sparkle through the waxed leaves of salal. Who wouldn't give up a chrome and tile bathroom for this?

The next morning we stood on the rocky ledge and stared at Bertha. Drawn out by the tide, she was nearly in the center of the cove. How small and lonely and fragile she looked, for our only link to the outer world. She was part of us now, part of the island.

"Dave, we'll have to build a dock," Jeanne said, reading my mind as usual. "Then we won't have to worry about her."

When we ferried over our possessions, we had discovered it was hard on the boat and us too, beaching it in the mud or on sharp-barnacled rocks at low tide. It wasn't practical to unload Bertha anywhere except at a float that rose and fell with the tides.

After scrounging more drift lumber and three husky beach logs, I started to work on the dock while Jeanne planted the garden. The deepest spot in the cove near shore was off a steep sandstone bluff fully three hundred yards from the shack and still farther from the prospective cottage sites. It would force us to do a lot more packing and wheelbarrowing, as well as to grub out a long path through the dense willows and madronas bordering the cove. But we were young and strong. Nothing seemed impossible. Before long we hoped to get a secondhand tractor and also to bring over the trailer, to do the hauling for us. Little did we know that we were to wait three years for a tractor and that more time would be spent getting materials to a cottage site than building a cottage.

While I banged away at the float—measuring, cutting, and spiking eight-foot stringers across the logs—I was never alone. Kingfishers dove into the water and darted back to their fishing limbs. Black and white buffleheads paddled curiously around me, chattering guttural comments as they watched my weird creation take shape. Even though my hands hurt, my knees were raw, and my shoulders ached from swinging a hammer, I didn't mind. I was a boy building his first raft, too excited to care. For I loved every aching minute of it; I loved the feel of driving a long spike home, the taste of sweat on my lips, the smell of freshly sawed wood. I loved listening to the fluted call of the eagle, the shrill cries of gulls, and the swish of ducks scooting onto the cove. Best of all, I loved to rest on my haunches and puff on a cigarette, letting my eyes wander, knowing that wherever I looked I could say: This is all mine.

I found, as I toiled, no greater thrill than that of learning a fresh fact and putting it to use the same instant. By blunting the head of a nail, I discovered, I could drive it in without splitting the board. And to pull out a bent spike, a block under the hammer claws worked wonders, saving nails and hammer handles as well. Yet learning to be your own naval architect is a tricky proposition. Your hands are always so wet that it's a job to hang onto your tools—and nothing is more maddening than trying to locate a hammer on a dark, weedy bottom.

Even while you sleep, the sea is working against you. If it doesn't claim your tools, it'll do its best to wreck them. Rust obliterates the figures on the steel rule and the square. The saw begins to act like a balky child, and the ax has suddenly lost its appetite for wood. Before long, you can't even get your zipper down when nature starts to call.

The questions of buoyancy and balance plagued me. They applied as much to me as they did to the float. After my first two unexpected dips, weighted down with tools and pocketfuls of nails, I realized that I couldn't depend on coming up automatically. From then on, my balance miraculously improved.

The logs were not too buoyant either. They were firs, and should have been cedars. As each timber went on, the float went down. Water spilled over my shoes. I couldn't feel or tell where I was stepping. I swore never to use drift logs again and to build the next float on the beach.

When the rickety little wharf was finished a few days later, another problem faced me. The float faced the open gap. A strong wind might smash it against the sandstone wall. The weight alone of the sixteen-foot gangway of fir saplings and drift boards, my primitive engineering brain told me, was not to be trusted to hold it. The top of the gangway rested high on the rounded bluff, roped to a squatty fir; its blunt legs sat perpendicularly on the float. When the tide was out, it looked for all the world like a ramp running up the side of a whale.

Then an idea came to me. By noon, I had secured the float by wire to four huge rock anchors, dropped fifty feet off each corner. I stood admiring my ingenuity, stripped to the waist in the warm sunshine. The summer stillness of the bay gave no hint of what winter winds held in store.

We christened the harbor Conover Cove. "One thing for sure," Jeanne said as we docked Bertha against the float, "we'll not be bothered by traveling salesmen."

There was seldom an idle moment between daylight and dark. If we weren't hacking out the trail to the float, or beating back the jungle of firs and willows that threatened the orchard, or pruning the fruit trees, we were digging in the garden, collecting drift lumber, salvaging boards and nails from the old camp buildings, or constructing the chicken house. Our hands were seldom empty, our backs always bent. Blisters refused to heal, and calluses came slowly and painfully. We called them "medals of honor above and beyond the forty-hour week." Night found us numb from exhaustion—it seemed as if we'd never adapt ourselves to so much physical labor. But even working sixteen hours a day and often by lamplight, we never accomplished all we set out to do. "Beat you to the salt chuck!" announced moments of play, supervised by crows gawking noisily from the madronas and gulls looping overhead.

We were lost in a timeless world. The days rolled by, nameless and never the same, each crammed with the joy of living. Working with our hands seemed to give us an inner contentment we had never known before. A sense of peace and humility enveloped us as the island whittled us down to size. How deplorably ignorant we were. How barren spiritually. We found ourselves drawn to the Bible, and began reading a chapter each night, a habit we've never broken.

As the middle of June approached, we were two people bottled up in a dream. We dreaded the day we would have to leave our paradise to make a shopping trip to the village.

CHAPTER 6

Noah's Ark

NO TWO CRUSOES EVER HAD IT SO GOOD.

Treasures landed on our doorstep every day: wicker birdcages and glass balls from Japan, boxes, shoes, two-by-sixes, grapefruits, even bolts, pots, and pumps. Name it, and eventually it would arrive. Of all the pleasures of island living, beachcombing is by far the most exciting and rewarding pastime. A form of privileged lawlessness, it satisfies that urge to steal, to get something for nothing. And it arouses a sense of adventure and wonder, since you never know what the tide will bring.

"If I didn't have housekeeping to worry about, I'd be out scrounging all day," Jeanne confessed.

She was the eagerest beachcomber you ever saw—a sight to see in shorts, work gloves, and a frayed straw hat, scrambling along the rocky shores and tossing up prize drift. Not a day passed that she didn't find something unusual: a leather purse, cap, flashlight, false teeth, or a fishing plug. She'd come back from each quest with her arms laden, her face beaming as though she had just bought a new dress.

Every morning we rushed outside to see how much the garden had grown overnight. We nibbled on young shoots of parsley and lettuce, pointed with pride to rows of radishes, carrots, and beans, which were outracing the weeds. All that back-breaking work removing the tons of rocks and roots suddenly seemed worthwhile. Unfortunately, the soil was light and sandy and demanded water nearly every day. Bringing it down by the bucketfuls from the well was a tedious task, and watching the earth swallow it like a sponge made me long for the day when we'd have water piped to the spot. I liked the feel of the soil and of growing things. It unlocked the lion in me. Everywhere else Nature defied me, but here in this twenty-by-forty garden patch, I was king.

Jeanne was far more practical. To her, the garden was a money-saving proposition that, come mealtime, paid dividends in convenience and taste. However, direct contact with the soil was something to be avoided. She made no bones about it. Elbow-length gloves always guarded her pretty hands, and when a bit of dirt slipped inside, she'd rush off to wash her hands.

We were grubbing weeds one morning when a small sailboat with a Canadian ensign chugged into the cove. A blond man, bare to the waist, stood at the tiller.

The sight of a visitor suddenly made us feel lonely. Jeanne reached for her blouse, and then we started running toward the float, waving our arms, as the boat crept closer. We could see something small, brown, and hairy, like a monkey, climbing up the mast.

"Ahoy there!" How good those words sounded. "Is it all right to dock?"

We shouted our approval.

The sailboat nudged against the float. From nowhere, little brown puppies appeared and waddled along the sloping deck. Though the portholes, an odd chorus of chirping and twittering came to our ears. What kind of boat was this, I wondered. Just then a gray kitten scampered onto the cabin roof, sat down and stared soulfully at us.

"Isn't he cute," Jeanne exclaimed. The smiling young Adonis was slipping into a shirt as he stepped from the cockpit. "Sorry, it's a she," he said, and he put the kitten into Jeanne's willing hands.

"What's *her* name?"

"She hasn't one yet—or a home." Then he introduced himself, and his handshake was warm and friendly. Colin Hanney had a quick, winning smile. "Would you like to buy her?"

Jeanne pressed the kitten against her cheek and looked at me hopefully. We both nodded.

"I should have told you," Mr. Hanney went on, eyeing his boat with pride, "this is a floating pet shop. I have a store in Vancouver—the Noah's

Ark. Peddling pets and supplies is the way I pay for my summer holiday. Come aboard and see my menagerie."

We were flabbergasted. Our first visitor was, after all, a salesman. But we liked Colin immediately. His Nordic blue eyes met yours squarely, and he had the easy, unaffected manner of a young man who knew what he wanted from life.

It was astonishing how much his cramped little cabin contained. Budgies and canaries hung in cages from the ceiling, and the floor was littered with cartons full of puppies of various sizes and colors. Dog and cat food lined open shelves against the hull, along with leashes, bird baths, and fish bowls. Packages of canary seed and worm tablets were stacked on his bunk.

Colin removed a floorboard at our feet and pointed to a tank in the bilge. "Some of my stock is here." Goldfish darted to and fro, and turtles scuttled out of sight.

Our ears throbbed from the din of voices crying for attention. To live with all this clamor and confusion, we were convinced Colin must have an extraordinary liking for his pets. We wanted to know him better.

That evening he came to dinner. He looked a little surprised at our simple quarters, but admired the spray of wildflowers and the fresh tablecloth.

"Hope you didn't go to any bother," he said as he sat down.

Jeanne blushed. She had fussed all afternoon.

"You don't look like the wealthy Americans everybody in Ganges is talking about," Colin said later. "I'd say you're just ordinary people."

"That's odd"—Jeanne glanced at me—"nobody knows we're here except Vic."

Colin grinned. "Not much, they don't know about you. The odds are ten to one you won't last the winter. They think you're crazy—but I don't think so. With an island like this, I think you're the luckiest kids I've ever met."

We had much in common. His hobbies were mountain climbing, boating, and photography. We climbed Mount Rainier all through dinner and discussed lenses, cameras, and boats until the oyster shells were filled with cigarette stubs. We wondered if all young Canadians were like him—he was no different from our friends at home. Nor were we the gregarious, dollar-happy souls he had pictured Americans to be. Wherever Americans go, they have to buck this picture of themselves. They may be the world's best salesmen, but they have failed miserably to sell themselves.

We hated to see Colin leave the next day. Yet we were sure he took with him almost as much as he left behind. As his boat slipped from the cove, I turned to Jeanne. "You know, there's always something about a man who loves animals that makes him worth knowing."

The kitten followed us around like a dog. We tried different names, but they fell on deaf ears. Then Jeanne said Kiki. The kitten stopped in its tracks and twisted up its head as if to ask, What do you want? The little scamp finally had a name.

The rats were so large we were afraid they might kill Kiki. But, shortly after midnight, we felt something bounce on our bed. I grabbed the flashlight. There was Kiki, meowing boastfully, with a rat the size of a hank of bologna between her paws, waiting for us to praise her. After several nights, we tired of this routine and closed the window.

In her diary, Jeanne wrote of Kiki:

> Kiki is dreadfully sober and unkittenish, with her tigerish expression and little hornlike ears. Can't stand to be alone. She loves oysters, and is a perfect nuisance when we dig clams. Our crow friend Freddie likes to tease her, and she stalks him ferociously—shoots up the tallest fir after him, then sits there calmly and smugly admiring the view. She doesn't mind the boat—the noise of the engine or getting her feet wet. In fact, she's something like Dave—nothing bothers her.

Wallace continued to awe us with its beauty. The sword ferns were head-high now, the ravines so chock full of wild, tangled greenness that it blotted out the sun. Delicate ladyslipper orchids hid their orchid-blue beneath the salal. In contrast, vines hung from the firs thick and strong enough to swing on Tarzan-style. The drone of insects and the scent of wild roses were everywhere. Lizards basked on the warm rocks as on a tropical island. The marshes rang with the music of yellow canaries, towhees, and larks.

We began to see more pleasure boats, but the reefs kept them a half-mile offshore. That didn't hurt our feelings—we had no time to spare for visitors, and it was so hot we wore few, if any, clothes.

The next day was the twenty-fifth of June, but I had forgotten my birthday. When I smelled something burning, I raced to the shack in alarm. Jeanne was in tears beside the stove. On the open oven door lay a black and smoldering cake.

"I wanted to surprise you, darling."

I folded her in my arms. "Nothing comes easy on an island, does it?"

Dinner that night made up for the cake—fried oysters, biscuits, wild strawberry jam, and applesauce topped with whipped canned milk. When we finished eating, Jeanne said, "Close your eyes—don't peek!"

I heard her put something on the table.

"Okay, you can look now."

There, under a flutter of ribbons, was a barometer. It became an oracle to be tapped each day for a clue to the weather, though later on, when Vic heard about it, he jeered, "All th' bugger will do is scare yuh to death."

We sat up till midnight going over the shopping list. It looked as if we were setting up a camp on the moon: spade, blowtorch, drum of gas, 200 BM two-by-four's, chickens, 600 feet of iron pipe, a case of canned cow, kegs of nails, cement, seed potatoes, wheelbarrow, and so on.

"We can't hold out much longer," I said, studying the list. "Maybe we'd better go in to Ganges tomorrow."

"And let's not forget to put up the mailbox at Fernwood," Jeanne said.

The chicken house had exhausted our supply of good lumber. Now we were ready to start on a storage shed—a place for firewood, building materials, and other supplies—so that neither weather nor shortages would interfere with our goal: a home and a guest cottage by next summer.

We were up and dressed for town early the next morning, looking like dudes. I shaved and put on clean suntans, and Jeanne wore green slacks and matching oxfords. She stood in front of the mirror wriggling her shoulders, concerned at the snug fit of her fresh blouse.

"Do I look okay?" Her voice was doubtful. "We've got to make a good impression, you know."

"You look great to me."

And indeed she did. The white blouse set off her golden skin. Her figure was lean and compact from hard work, and her green eyes sparkled with happiness.

When we reached the float, she turned to me hesitantly, as if something was bothering her. "Maybe we'd better wait, Dave."

"Why? It's a perfect day to haul our stuff back."

"I can't explain it. Just a hunch that we shouldn't leave the island."

"Don't be silly—nobody's going to run off with it. Let's get going."

Before long we were in the car bounding over the potholes toward Ganges. A harbor nestling a log boom sprang into view through the trees, and a shack with the words BARBER SHOP shot by the window. Then a log-cabin cafe and gas station popped up on our right; a boat repair shop leaned on its pilings to the left.

"Reminds me of a ghost town," Jeanne said.

We drove past the blacksmith's shed out onto a bleak bite of land where three or four weary gray buildings huddled about a large wharf, like remnants of some long-ago disaster. On the most impressive clapboard structure, a faded sign read: MANDER & SONS GENERAL MERCHANTS. A row of ancient cars nuzzled the store like piglets suckling a sow. We nosed in between two sporting running boards, and Roberta wheezed contentedly to a stop.

The store was a two-story education in country living, though we were doubtful of just what country. Permeated with the smell of rope, coffee, and rubber boots, the atmosphere was one of mañana. If you weren't in a hurry (a vulgar word here), it sold everything from marriage licenses to coffins, tea cozies to Ford station wagons. Money was handled by an overhead monorail system that shuttled a cashbox through walls, along shelves of liniment, catsup, and candy in glass jars, and under lanterns, pails, and feedbags hanging from the ceiling. A Coke dispenser squatted at the soda fountain; "Player's Please" sang out from the walls, and groceries greeted us in French.

"What's the French for?" Jeanne eyed the box of cornflakes. "Didn't General Wolfe defeat Montcalm?"

"Maybe so, but he must have lost the war."

We felt foreign, too. A crowded bench of old-timers abruptly stopped mumbling and returned our gaze with solemn stares.

"I'd feel better in my dirty denims," Jeanne whispered.

Our clothes were not impressive, but compared to the loggers and farmers, we stuck out like sheiks. People made their purchases in hushed monotones, agog with interest as we wandered through the store to the post office in the rear.

"Thought you'd show up, Mr. Conover," the postmaster said with a grin. "Vic said you'd be in." He shoved a bundle under the wicket and assured us that our mail would be delivered to Fernwood. "Don't wish you younguns any bad luck, but I've got a twelve-to-one bet you won't last the winter."

Tempted by the odds, I was just about to get in on the betting when Jeanne yanked my arm. "Come on, we've got shopping to do."

Groceries were kept strictly out of reach behind long oak counters, on shelves crammed to the ceiling. The only "self-serve" items were fruits and vegetable in crates thrown open on the floor. Elderly clerks in smocks hustled about with a somnambulistic air of busyness. When one of them found the ladder, spent twenty minutes locating a can of spinach on the top shelf (which ran the length of the store), and then—out of breath—brought it down, we felt compelled to buy it.

The prices were sky-high. "At forty-two cents a pack," I grumbled, "who can afford to smoke?"

"What's the monkey cage?" Jeanne asked, nodding at the open, jail-like room full of groceries in the center of the store.

"The cash and carry, sah," the clerk replied Britishly.

To be seen inside the cage, we soon discovered, was not good. Either one's credit had expired or he was broke. Cash was used only by strangers.

The butchershop consisted of white-enamel trays of meat, salted fish, and hovering flies. A blackboard on the wall gave the prices.

"Hamburger, seventy-five cents!" Jeanne sputtered. "You don't know it, but you've just become a vegetarian."

"That's nothing. Look at this. I walked over to a wheelbarrow that I had seen at Sears for $9.95. "Nearly twenty bucks." We were dismayed at what these prices would do to our meager funds.

A lean, sober-faced Englishman called the Deacon presided over the hardware section like a custodian of the crown ewels. His smock was impeccably ironed. His spaniel eyes sagged lower each time he informed us he couldn't oblige, which was every other item on the list. He painfully counted out screws as if they were silver dollars. We needed brass screws and galvanized nails for shingling. We settled for iron screws and iron nails. For the water system we wanted plastic pipe.

"Plastic pipe! What's that?" We settled for iron pipe.

"Any elbows or nipples?"

He shook his head. "Be in any day now."

We hurried outside. The noon sun was warm, and the fresh air felt good.

"Let's pick up the building materials," I said. "Then we can get out of here." We were too depressed to think of food.

The lumberyard proved to be a pile of split, knotty boards on the wharf. My spirits hit a new low.

"No two-by-four's?" I growled at the man.

He shook his head. "Be in any day now."

If we heard this once more, we would scream. We took what he had, and all the shingles and cement we could carry. Then, discouraged, we limped from town with pipe strapped to the fenders and lumber dragging in the rear.

"With this load," Jeanne sighed, balancing a wheelbarrow on her lap, "we shouldn't have to come back in months." I hoped she was right.

As we drove toward Fernwood, it was difficult to shake off the mood of Ganges. It seemed to us like a stranded settlement that had died long ago; everything about it was strange and unreal. Even its smell had a mustiness, an odor of decay like sewage wastes at low tide. The moan of a distant chain saw magnified the deathlike stillness of the place. The only thing we would ever have in common with the people there, we thought gloomily, was the same tongue. Later on, closer acquaintance erased this impression. British blood prevents easy friendship. Shy and polite, the townspeople warm up slowly and are reluctant to accept strangers, particularly Americans who do not live on their island.

Not until we saw Vic chasing after chickens with a fish net, so that we could take them back to the island, did we feel like smiling. His fowl were as wild as his cats, and his yard was an obstacle course of old toilet bowls, hot-water tanks, and rusty machinery hidden in tall grass. He cornered the last one under his expired Essex.

"Hold onto the sack till yuh get 'em locked up," he puffed, spitting feathers from his mouth.

Holding onto the sacks of groceries was a different matter. As we loaded the boat, they split open and rained canned goods into the sea. Sacks fell over and broke eggs, squashed tomatoes and cookies. Sacks made us sticky and dirty; the gangway, slippery and dangerous. Sacks got soggy, dribbled, and fell apart in our arms. Sacks! Sacks!

"Let's get boxes after this!" Jeanne exploded.

When the cove drew apart in the trees, we thought nothing in this world could be lovelier. The last seven hours seemed like seven weeks. No sooner had the points closed around us than we noticed a scruffy-looking fishboat anchored near shore. Burning with curiosity, we circled it but could see no one. Whoever the visitor was, apparently he was ashore.

Kiki was pacing the float, meowing excitedly as we pulled alongside. Not waiting to unload, we marched hurriedly up the path. Kiki preferred to stay behind. Halfway to the shack, we confronted a dark-skinned man strutting down the path. A dirty felt hat shaded his eyes, and his patched trousers clung sensuously to his legs and thighs. Jeanne's grip tightened on my arm. The way he walked sent a chill through me—the swagger, the half-wild glint in his eyes, the sphinxlike expression.

He stopped in front of us. "Got water?"

For an instant our tongues were tied. We faced a fullblooded Indian. We mumbled Yes and told him to follow us. Jeanne clung to my arm and peered nervously over her shoulder now and then as we walked on toward the shack. I found a pail and filled it for him from the tub near the garden, and he went back to his boat.

"Odd, isn't it," Jeanne said. "He didn't even have a container. Suppose he was casing the place?"

So often Jeanne had an uncanny way of going right to the heart of a matter. He was up to something all right. His eyes were shifty, the kind that wouldn't look you straight in the face.

"You keep an eye on him," she said, heading toward the float, "and I'll start bringing things up."

I went down to the beach and collected the pail from the Indian. After some difficulty starting his engine, he departed noisily from the cove. The next moment I heard a commotion at the wharf. Jeanne was sprinting up the ramp after Kiki, who was two short breaths behind the terrified, squawking chickens hell-bent for the woods.

It was almost dark before we recaptured our flock. Then the cause of it all rubbed against our legs, purring delightedly in appreciation of the fun she'd had, as we groped, tired and hungry, toward the shack.

"Put on that hamburger," I ordered. "I'm in no mood to be a vegetarian tonight."

For punishment, we made Kiki wait for her bread and milk. She did her best to disrupt the cooking, complaining bitterly of her hunger, wandering between our legs. We held our tempers, only to find her on the counter protesting her need more loudly. This had always worked before. Finally she slithered under the stove and became a woolly round ball.

We ate dinner with little relish; our minds were not on our food. When we had finished and sat smoking, Jeanne said, "Dave, I don't think we should leave the island alone again. Do you?"

I shook my head glumly. Our Eden had suddenly produced a serpent. Theft was the last thing in the world we had expected to worry about. We decided, because of the many heavy items that had to be brought over besides food, that I would do the shopping from now on.

"Won't you be afraid of being here alone?" I asked Jeanne.

"Not as long as I have a gun."

Later that night I dug out the .22 rifle and hung it on a hook near the door. "There now," I said, after explaining how it worked, "we're ready for business."

Secretly, I felt a bit safer too. With a nail across the door latch, we went to bed a little sad, feeling that the intruder had already robbed us of something precious.

CHAPTER 7

Brown Water

SEVERAL DAYS LATER, WHILE I WAS LOADING SHINGLES at Fernwood, a freight truck roared down the hill and dumped a pile of two-by-four's at the head of the wharf. I groaned. The grim task of another day's hauling was more than I could face.

"We'll raft it," I said to Jeanne, who had come along to help with the last load. "If we hurry, we can make it before dark."

She eyed the pile doubtfully. "Can Bertha pull all that?"

I checked the tidebook. "Don't see why not. We've got the current with us for a couple of hours, and the sea is flat."

We carted and dragged the lumber down onto the float, tied it into a large bundle, and shoved it over the side. For the first half-hour, we made good time. Halfway across, the sun dropped behind Salt Spring and a north wind began to freshen.

Jeanne peered at Wallace's dusky shore, then at me. "Something's wrong, Dave. We're not getting anywhere. Did you read the tidebook correctly?"

"Of course."

"Did you take off an hour for daylight time?"

I admitted sheepishly that I had forgotten. The tide had turned twenty minutes ago.

Shoving the throttle wide open, I dropped down at the tiller beside her. We huddled there, bending our heads each time the spray swept over the bow.

"Shouldn't we turn back? It's getting worse."

"Can't," I answered. "We'll lose the lumber—it's only tied with a slip knot. Just have to sweat it out."

The waves grew uglier. Thundering now, they battered the bow and exploded into stinging spray that chilled us through. Before long, the black shadow of the island dissolved in the dark. We were still a half-mile off Panther Point.

"We're holding our own," I said, trying to be cheerful. This was Jeanne's first tight squeeze, and I didn't want her to become afraid of the sea.

"How can you tell? *I* can't even make out Wallace."

"By Vic's light. It's still in the same spot. And, I thought, it better not go out, for the moon wasn't due to rise for another hour. A lot could happen if we wandered off course. Panther Point reefs extended a quarter-mile off the tip of the island. The coal ship *Panther* sank on them.

Bucking into the charging seas, Bertha groaned on. My hand grew numb on the tiller. Both of us were wet and shivering. We could see the phosphorescence boil over the lumber as I kept Vic's light over the stern. All the time I was wondering if we had enough gas and hoping the tide would hurry and turn.

The next hour or so we hovered somewhere off the island. The blackness added fury to the striking waves. White lips snarled by angrily, seething into the dark. The smell of evergreen grew stronger, the exhaust boomed louder. When the moon rose, tall masts of firs crowded the bow. I swung hard to port, and the lumber thrust ahead, barely missing the jagged rocks. Toward midnight the tide slackened with the wind. We spied the cove's silvery outline and turned into the gap.

Suddenly Bertha shot ahead.

"We've lost the raft," Jeanne cried.

By the time we circled back, boards were everywhere.

For hours we salvaged lumber. What was worse, the float was dry and we couldn't dock. Disgusted and weary, we waded through the mud with the boards while Kiki sat mournfully on the float and supervised the piling. Later, Jeanne brought down coffee and cookies.

"Think we'll lose much?" she asked.

"Not a board will get away," I promised. "We've got the tide on our side for a change."

At daybreak, we ate breakfast along with the dinner we had missed the night before. As the first squadron of crows took off for Salt Spring, we fell into bed exhausted. In late afternoon thunder awoke us. Water was dripping on the bed. Rain blew in the windows and trickled down the walls. We rushed every pot and pan into use and wadded newspapers into the largest cracks. We even built towel dikes around the door.

"Now I can wash my hair," Jeanne announced cheerfully.

Water plinked and plunked into pans while we raced frantically about emptying them. Chunks of soggy moss fell on the floor. Now I knew what had been covering the leaks. Twice that night, in the lashing rain, we stumbled to the wharf to check Bertha's moorings.

The next day the rain stopped and the mosquitoes came. The hot sun drove them out of the forest with a vengeance. We sweltered in long pants and long-sleeved shirts, but still we began to look as if we had measles.

Work on the water system had to be postponed, so I climbed a ladder of fir poles to shingle the roof. Between the slippery moss and the wet shingles, it was like clinging to the slopes of Mount Everest. The book said, without benefit of drawings or pictures, to start from the eaves. I hollered to Jeanne to look up "eaves" in the dictionary. Then the next sentence struck me: "Amount of lap is governed by the pitch of the roof." My mind balked. How the devil would I know that! Disgusted, I threw the book to the ground.

"Now what's the matter?" Jeanne stood below, wiping her head with a towel.

"I guess I'm too thick. I'll just have to follow the old shingles."

The next few days I was a novice trapeze artist. Crawling about on all fours, I never knew whether I was going to go through the roof or fall off it. I swatted more mosquitoes than nails. Shingles kept splitting and falling, ramming splinters into my hands and knees. The nails were too big. The hammer kept dropping off the roof. Finally, I slipped and bounced off into the apple tree, my dignity more hurt than anything else.

Jeanne rushed outside. With true feminine logic, she snapped, "Can't you be more careful? How do you expect the bread to rise?"

Dinner that night was still a success. Jeanne passed the test of a true pioneer with her homemade bread, clam chowder, and wild strawberry shortcake. I never felt more pleasantly stoked.

"For this meal you deserve something special," I said, digging out a chrome tap from a box on the floor. "Now that we've got pipe, you're going to have running water."

But I had no idea how to do it. The old well was in a thickly wooded ravine three hundred yards off, and the water level—as I could accurately testify—fully twenty-six feet below the surface. When the dishes were done, I mustered all my unborn surveyor's instinct and cut a path through the bush, charting as straight a line as possible between the two points. That night, I studied everything I could find on water systems. In an hour, I was more confused than enlightened.

I showed Jeanne what I was reading. "*Plumbing Simplified!* That's a laugh." Tossing it aside, I reached for another. "*How to Get Running Water to Your House*—doesn't that sound great? The guy who wrote that never lived on an island." Nothing was possible, it seemed, without electricity.

Jeanne was busy writing in her diary.

> JULY 6: The roof is finished, thank heaven! All day it was, "Toss me the hammer. Get me those shingles. Don't throw them, stupid! If you aren't doing anything, bring up the nails. Did you check the chickens? Hand me the ladder. Long as you're here, hold this board. Not there! Can't you do anything right?" What a day I picked to bake bread!
>
> Kiki woke us up last night growling like a cougar. We dashed outside. There she was, squatting in the lettuce, as a crash of hoofs echoed in the bush. Not a thing was touched. We'll never need a watch dog. Rewarded her with all the oysters she could eat.
>
> This morning Dave tinkered with the broken door latch for hours. His patience irritates me. He's not satisfied to get something working—it must sound right too. He keeps after me to try the boat alone. I hate to tell him, but after that scare the other night, I just haven't the nerve.
>
> Not one egg yet. Chalk it up to strangeness, I suppose. I located one of Dave's golf balls and tucked it in a nest. This afternoon I found them kicking it about the yard. Dave says maybe all they need is a rooster.

He'll have to go to Victoria soon. Ganges hasn't half what we need—doors, windows, plywood, bricks, finishing lumber, etc.—and the prices are terrible. Necessities are luxuries here. A gas refrigerator is $500. We're doomed to drink Jello the rest of our lives. I'm depressed tonight. How we'll manage, Lord only knows!

Sleep that night was a plumber's nightmare. I stood on a mountaintop of pipe, battling off winged pipe wrenches, hacksaws, and plumbing books. Surprisingly enough, a scheme did present itself the next morning.

"With an airtight pipeline," I told Jeanne after breakfast, "the water will run to the house by gravity."

She went on washing the frying pan. "Don't be silly, Dave. How can you drag water out of that deep hole and get it here without a pump?"

"Wait and see."

To prove my theory practical while my enthusiasm bubbled high, I joined the pipe on top of the ground to try it. Then I primed the line at the well. Barely able to contain my curiosity, I routed Jeanne from the garden and we rushed in the door.

"What do you bet?" My hand grasped the tap.

She shook her head.

I turned it. Nothing happened—not even a wheeze.

My spirits collapsed like an accordion. Any moment I expected to hear "I told you so." Instead, she said, "Now that we're here, we might as well act like Canadians and have a cup of tea. The kettle is still warm."

I sank into the chair, wondering what was wrong. A leaky joint? No, I had checked each one twice. As Jeanne set the cups on the table, the tap gurgled. Water shot into the sink.

"Holy cow! It works." She looked at me approvingly. "I'll do a wash right now." And she began filling pots to heat water.

Happily, I withdrew and commenced to dig the pipeline. Roots, rocks, and trees! I wielded an ax more than a shovel. Digging was the worst part of plumbing—and a trench through the bush, an absolute back-breaker. Sweat stung my eyes. Every time the pick whanged a rock, I felt lightning strike me. In a couple of hours my hands were raw and bleeding. I retreated to the shack—to find Jeanne huddled over the tub, on the verge of tears.

"Look at our white bath towels," she sputtered, holding one up to show me the dirty brown stains. "They're ruined. What's wrong with the water?"

We spent the evening gloomily combing through books for an answer. Time after time, we ran the tap and held a glass up to the lantern. No rust, dirt, or peculiar taste to give us a clue.

"We can't run a resort with this water," she fretted, "let alone wash a sheet."

The next morning another headache greeted us. No water at all. It had drained back into the well. Priming the line every day was far too laborious a task. For an hour, I thought my system a failure. Then, to my relief, the plumbing books disclosed I had forgotten a one-way valve.

"Maybe Vic has one," Jeanne suggested. "His place is loaded with old junk. And you could ask him about the water."

Fortunately he found one in a box of pipe fittings under the barn. "Do a little plumbin' on the side," he confided, brushing the dirt from his beard. "Relievin' septic tanks and drains mostly, though . . . . Yuh look worried, Tiny. What's yer trouble?"

I told him.

"Iron." His lips puckered and a puff of dirt rose from the ground. "That's th' curse of sandstone, my boy. Yuh can't do a blame thing about it."

"You sure?"

He nodded. "Yuh jest gotta live with it."

Maybe so, I thought, but I knew Jeanne couldn't. She'd soon grow to hate the island. Realist that she is, a woman will never gamble on a murky future, where a man always will.

With a rooster, an old range boiler, and coil, I headed for home, wondering how I could lick the problem. At least, now we knew the villain. But water-softening equipment, which we had read about, was far too costly and impractical. I had to find some way to remove the iron simply and inexpensively.

My experiments turned the kitchen into a mess. For hours Jeanne watched curiously as I rigged up various filters in the sink: sawdust, clamshell, earth, leaves, charcoal from the stove, even a pair of nylons. After each experiment, we washed a piece of white cloth to see if it would turn brown. It did every time.

On Friday we had hard-boiled eggs for lunch. The pan of water they had boiled in sat all afternoon on the cold stove. When I was repairing the oven door late that day, I noticed a thin layer of brown sediment on the bottom of the pan. My curiosity was aroused. I quickly poured off the water, washed a white handkerchief, and held it up to the window. It was still white!

I ran into the garden.

"You mean if we *boil* water," Jeanne repeated unbelievingly, "the iron will settle?" She stared in amazement at the wet handkerchief.

"Seems to, when it cools. Exposed to the air, it turns into a brown sludge."

"But what an awful process to go through to wash clothes. And what about summer guests? We can't tell them you've got to boil the water and drain off the sludge. We'll just have to find some other way."

That was all I could think about the next day while I buried the pipe—another way to expose the water to air. The solution was so simple I was overlooking it for something more difficult. At dinner, Jeanne remarked, "Why don't you spray it?"

In no time I was on the roof, aiming the hose down at the washtub on the ground. There was just enough pressure to give a fine spray.

Jeanne remained below, where she could see the tub. Soon she was shaking her head. "No sludge or sediment—not a sign."

With long faces, we turned away from the tub and went inside. "Never mind," she said while we did the dishes. "Tomorrow, why don't you tackle the hot water? Then we won't have pots and pans all over the stove." A tub bath took every one of them.

Jeanne bathed with undying regularity. I asked her one night how this shameful bit of conformity had crept into her nature. She said simply, "So you'll want me more often."

Jeanne's honesty never won her many girl friends. To please anyone, she never said what she didn't believe. Nor could she fib with any conviction, for she always blushed. Even if it hurt, she never lacked the courage to point out my mistakes. This candor is a great source of strength to me.

I have a doggedly one-track mind; hers carries the overall picture. She quickly sees where a change in plan would help. By comparison, I am a horse with blinkers. She weighs things by dollars and time; I weigh them by permanence and looks. She wants results; I can wait. When I want something, she asks, Do we need it? Her face tells me everything; I hide behind mine. An odd partnership—but it works.

The hot-water boiler had stood in the corner for days, glaring at me every time I came in. As Vic's instructions echoed in my mind, I leaped into the job, feeling, for once, that I faced something easy.

Jeanne watched while I took the firebox apart. I handled the iron plates as if they were gold. Shortages plagued me. No 45-degree elbows! Instead, we bent the galvanized pipe in the crotch of a madrona tree. Finally, when the kitchen corner looked like an oil refinery and the walls had ugly holes, we fired up Bessie. I felt success surge in my veins.

Whoosh! A great gush of steam billowed from the tap. The boiler banged and danced on its spindly legs. Kiki stared at it wide-eyed as we retreated toward the door.

Jeanne grabbed my arm. "Will it explode?"

I wasn't too sure. "We'll just have to wait for Bessie to cool before I can look for the trouble."

My mistake was boldly apparent. I had reversed the cold and hot-water lines. "How stupid can I get!" I muttered as I ripped out the pipe.

Jeanne laughed. "Nothing comes easy on an island, does it?"

Later that evening, while I was studying carpentry, Jeanne announced that she was going to take a bath. She faded behind the curtain to undress, then asked me to bring in the tub.

When I set it on the floor, I couldn't believe my eyes. "Come quick," I called.

Jeanne rushed out.

"Look!" I pointed to a slimy coating of iron in the bottom of the tub.

"Holy cow! The spraying did it."

"Now," I gloated, "you'll have the whitest sheets of your life."

We whooped and danced around the tub for joy.

CHAPTER 8

Mr. & Mrs. Crusoe

BY MID-JULY, WE WERE READY TO TACKLE THE SHED. For convenience, I decided to build it against the privy, and I was determined it would be twenty-four by twelve, level and square, even if it killed me. At times, I'm sure Jeanne thought that it would.

"You can't stand on that wobbly affair," she called when the frame was nearly up. I had no choice. She watched apprehensively while I straddled the rafters and began nailing them down. They were twisted and bent by the sun. They split. They sprang at me, cracking my ankles and knees. I fought for balance and the most expressive oaths.

But not knowing what I was doing was worse. I must have come down the ladder a dozen times to refer to the plan. Finally, with a grin, Jeanne stuck it in my pocket. "You need it more up there."

Her remarks were not always encouraging. While I was putting on the sides, she protested, "You're not building the Taj Mahal, you know."

"These trees won't hide everything. It's got to look right," I argued.

"Makes no difference—you're wasting too much time and lumber. You've got to be more practical."

Perfectionism had been a curse of mine. I was never satisfied with myself or anything I ever did. Every photograph I took could have been better, every print sharper and clearer. But there was no time for perfectionism now. The island would throttle us first. I just had to learn to be content with my best.

The shed, however, was considerably less than my best. The roof rippled, the posts were crooked, and the open front yawned like a dark cavern. It had three windowless walls, a floor of flat rock, lumber racks, a workbench, and nails to hang tools on. But at last we had a place to put things. And we no longer needed to enter the privy so cautiously. Building the shed had also been good practice—for adding the lean-to bathroom onto the shack.

"For a bathroom," Jeanne declared that evening as I pored over the plumbing books, "I'd live in one room for the rest of my life."

The days settled into a blur of routine. Work began at seven. Jeanne groomed the shack, washed, then cared for the chickens. They were laying at last, thanks to Bennie, but not in the coop—they just sneered at nests. Gathering eggs was a daily treasure hunt.

The garden took up another big chunk of Jeanne's time. Things kept on growing like mad. The corn was five feet tall now. The wild roses had turned into pests, always springing up in the lettuce, and firs sprouted around the clearing, ready to reclaim their lost territory. She had to weed every day to keep the bush from taking over. And when the outside chores were done, she rushed inside to "preserve" peas and beans, struggling disgustedly with the leaky pressure cooker.

I had rigged up a sediment tank on the roof—the water sprayed into a drum, where the iron settled out. Unfortunately, it had a habit of overflowing. An entry from Jeanne's diary is revealing:

> Dave says the tank needs a float valve and that "we'll have to watch it." That got me. "That means I'll have to watch it," I reminded him. "Wouldn't Vic have a valve?"
>
> He really got mad then. "I'm tired of running over there! Look—let's run this ship ourselves for a change." I could boot him one.

For days, I worked doggedly on the hole for the septic tank, but the island resented both pick and shovel. Hardpan slowed progress to a crawl and gripped the tools like cement. My flesh wept. Boulders round as manhole covers bogged me down. My groans brought Jeanne out to help, and together we roped, seated, and wrestled them out. When it rained, the hole became a quagmire. I hated rain. It aggravated every job and drove us inside. After it stopped, we bailed and fought mud and mosquitoes. Then the sun burnt our backs fiercer than ever.

When we had toiled long and hard, and the heat became unbearable, we dove off the rocks and raced each other to the points. Then we lay on the little beach below the shack and let the sun bake us dry. The harder we worked, the more intense our joy in stealing moments of sweetness from each day.

The noon meal we often picnicked on the rocks: an Arab's lunch of bread, milk, dates, and cheese. (Vic always had milk ready when we picked up the mail.) Then it was back to the hole for me, to the garden for Jeanne. Sometimes we worked side by side for an hour—planting spuds or grubbing bush before we went back to our separate tasks.

I was the builder of our kingdom; Jeanne, the provider. How she loved to bring in booty! She discovered plum trees and Gravensteins, hillsides of blackberries, and at low tide she waded for crabs. She did her best to spear a flounder. The huge oysters she found—larger than my fist—were a struggle to open, but one was a feast.

"What will keep people from stealing them?" she asked me one day.

"Nothing. Just us."

We had no legal rights to stand on. Land between low and high tide belongs to everyone. We had to trust to luck no one would challenge us for the natural rights of our island. Unknown to us were those who wintered off the fat of the land.

The sea was truly Jeanne's oyster. When she had any free time in the afternoons, if she wasn't out looking for loot, she was diving off the rocks for scallops and sea urchins or collecting starfish to enrich the garden. She decorated our little room with exotic shells and glass float balls. Sometimes she was too adventurous, I thought. I made her promise that when I went to Ganges, she'd stay off the seashore. It held danger, not only from falling on slippery rocks, but from delicately balanced logs left by the retreating tide.

Dinnertime was an exciting adventure in seafoods: steamed oysters, crab cocktails, buttered clams, shrimp salads. Jeanne cooked the way she kept house; the results were excellent, but her method unpredictable. She rarely knew beforehand what she would prepare, but let the moment be her guide. In minutes, effortlessly (to my amazement), she could pull off a corking good meal. To her, a born innovator, a cookbook was never right. If a recipe required corn syrup, she'd use molasses; or lemons in place of vinegar. She often brightened salads with nasturtium leaves and nuts.

After dinner, we could hardly wait to get outside. Serenity possessed the island, like an extra dividend to the dreamlike days. We gloried in the long fragrant July evenings, feeling the warmth of companionship flow between us as we brought up more flagstones from the beach for the garden path, took pailfuls of clamshell to the chickens, or pushed the forest back from newly discovered fruit trees.

Twice a week we ran after the mail. We grew weary of newspapers—they were always the same: rumors of war, disasters, crime on the upsurge. We felt better without them. All the letters rang with the same questions: How are you making out? Aren't you awfully lonely? Whatever do you do for excitement?

"What a laugh," Jeanne said. "Nature runs a continuous show, twenty-four hours a day."

We had ringside seats to great robberies, watching the crows steal starfish dinners from dignified gulls; to life and death battles between ducks and hungry eagles; to the spectacle of fighting blackfish churning the channel into blood, thunder, and foam. No wide-screen drama could have thrilled us more.

At night, with the lantern glowing from the rafters, we mulled over the day's work, the future, the next project. Jeanne's common sense was a sounding board for my pet plans, though she knew little—or less—about the subject than I did.

"What's the cost of a septic tank?" She looked over my shoulder at the plan of the bathroom. "Sixty-five bucks."

"Wouldn't concrete be cheaper?"

"Yes, but it'll take a lot of sand—and time hauling it over."

"We've got more time than money," she sighed. "Besides, we'll need sand for the foundation, won't we?"

The next day, Vic told me where to find sand. "Down the road, Tiny—there's a whole bank of it. Yuh can't miss it."

I paid him for a dozen gunny sacks and started off. By late afternoon,

Roberta was groaning. Sacks filled the front seat, draped the fenders, and bulged from the turtleback. While I filled the last one, a crow chattered alarmingly from a nearby fir. I smiled, thinking it was one of my birds spying on me, until a dark shadow fell across the sandpit. Startled, I turned around.

An old man in tattered clothing, leaning on a cane, faced me. He looked as bent and brittle as a dry twig. Snowy hair curled down his neck, and his weathered face had the texture of an avocado. He didn't seem real, standing there, his hawklike eyes scouring me from under his outlandish hat.

"Aren't you the chap who bought Chivers' Island?" he croaked in a Scotch twang. Before I had a chance to answer, "Pretty young to be ownin' an islan', aren't ja, laddie?" He stared at my feet, and his cane waved at my loot. "What cha doin' there?"

"Taking sand over to the island," I replied, guiltily.

The old fellow snorted, as though my words had disturbed him. Then his eyes fastened on mine. "That islan' has sand, heaps of it."

"How do you know?"

He leaned on his cane and looked down, as if he was trying to scratch his memory. "Old Jack built a chimney once.... Found a bank of clay, fired his own bricks, and mixed the mortar from his own sand. Ain't nuthin' the islan' don't have, laddie."

I could hardly wait until he finished. "Where is it?"

"You've got young eyes," he snapped. "Keep lookin'." He started to go, then swung around. "Mind ja now, don't sell the islan' short again, laddie."

He sauntered off and disappeared around a bend in the road. I had forgotten even to ask his name.

When I was on the way to the wharf, Vic bellowed out the window, "Tea's on." I was sampling Myrt's fruit cake before I thought to ask who the old gent with the cane was.

Vic's blue eyes studied me soberly. "Nobody around here like that"—he glanced at Myrt—"is there?"

She shook her head. "You've been workin' too hard, Tiny. Why, you're as thin as a fencepost." Her voice held deep concern.

"Tain't good," Vic added. "Yuh begin seein' things."

"But I saw—"

"That's all right, son." His eyes sought the window. "She's cuttin' up a bit. Yuh better git. Jeanne'll be worrin' about yuh."

She was on the float to meet me. "What kept you so long? Dinner's been ready an hour." Her eyes fell on my wet clothes. "Was it rough?"

"Wouldn't want it any rougher"—I nodded at the sandbags—"with that ballast."

I told her about the old man.

"Strange, all right," she said, steadying the wheelbarrow while I pushed a load up the path. "But we've looked everywhere, Dave. There isn't any."

"Just the same, we'd better look again. I figure we need forty more sacks."

The next two days we searched the island on foot. Kiki tagged after us, wriggling through the bush, enjoying every sight and smell. When we reached Pebble Beach the next afternoon, we were tired and disillusioned. We sagged onto a drift log.

"The tide's out," Jeanne said. "How about a feed of oysters before we look around?"

"Okay. You start a fire, and I'll go back and get some catsup, coffee, and cookies."

"You and your cookies," she chuckled.

When I came back, Jeanne was pointing to the hole she had scooped in the gravel for the fire. "Sand!" she cried, beaming. "Loads of it."

Unbelieving, I dropped down and dug my fingers into it. Real sand! To make sure it was not a small isolated patch, we dug several more holes with a stick. Kiki dug one of her own. Sure enough, soft brown sand was everywhere—tons of it, a foot below the gravel. It even hid under the drift line that ringed the beach.

"The old guy was right," Jeanne declared. We were as excited as if we had found Chivers' gold.

After we downed the last baked oyster, we sat and talked by the fire in the twilight. The tide nibbled at the beach, spicing the air with a rich tang of seaweed. Ducks paddled the glassy surface. Farther out, a family of seals were teaching their youngsters to swim. Owls drifted from fir to fir, hooting softly, while kingfishers dove for their bedtime snack. On far-off rocks, cormorants stood like black statues drying their wings. We felt as though we were the only two people in the world.

When a sickle of moon climbed above the firs, I tucked Kiki under my arm and we started to go. On top of the drift, Jeanne looked back, her figure barely visible in the dusk.

"I hope I'm wrong," she said. "But I've got a hunch that finding sand came just too easy."

"Won't be easy getting it out. We'll have to pack it around by boat."

July slipped into August while we bagged sand and gravel. Lifting 75-pound bags in and out of Bertha and wheeling them up the path made my back complain, but I felt too happy not to have something hurt. The hauling was hard on Bertha, too. Wash from freighters pounded her on the beach, and like us, she was bruised and grimy from the task. Jeanne finally came up with the suggestion that we start building the forms, and haul sand only every other day.

Yet our drudgery had its rewards. There's a certain richness to days exposed to the sun and sea and the scent of evergreens from dawn until dark. It's like living in two worlds. The sea, where the bottom is so close to the top, quickens your responses; the wildness of the land sharpens your senses. As we grew more alert, the ordinary affairs of the sea, the forest, and the spirit became more exciting and meaningful. The forest lost its strangeness; each tree along the path and shore became a friend, and each majestic fir, a finger pointing toward God. The beauty around us mirrored His presence. We dreaded cutting a tree, often transplanting it instead. For we had begun to feel that we did not truly own Wallace. The island was His; we were only His stewards.

A few days later when I went over to Vic's after more sacks, Jeanne's premonition proved to be right.

"Yuh can't use that beach stuff," Vic drawled, pitching hay into his loft. "Jest ain't clean."

"Looks clean to me."

He drove his pitchfork into the ground and glared at me, his brow a storm of wrinkles. "That's not the point, son. It's too blame full of salt. To use it fer cement, yuh gotta wash it." My shoulders suddenly felt like lead. Wash it! We didn't have that much water—barely enough to get by on, in fact. What would it be next? Water we couldn't wash clothes in. Now sand that we couldn't use. Then the words of the old man came flooding back. Chivers did. Why couldn't we?

At dinner Jeanne sensed something was bothering me. I finally had to tell her.

"I knew it, I knew it! When things come too easy on this island, there's bound to be trouble."

By now, I should have known her hunches were usually right. But nothing irritates a man more than to rely on a woman's intuition. Man is a great believer that all things are possible. A woman's life is spent reminding him they aren't.

"Wait a minute—didn't the old man say Chivers built a chimney?" Jeanne asked.

I looked at ours. It could be—but there was no way of telling if he had washed the sand for the mortar.

"Vic's been wrong before," I said. "We'll write to the cement company and find out for sure."

In the meantime, the mountain outside the window continued to grow.

The island was a great teacher. There is a keen sense of joy in acquiring new skills and disciplines, in learning the rhythm of the tide without consulting the little gray book, in understanding the whims of the weather and sea. Heavy dew meant a clear day ahead, and dark, low scurrying clouds meant an angry sea. Every day we learned something new—often without knowing it, and most often about ourselves. Man is a marvel of adaptability. Little comes along that he cannot either bear or change to suit himself. We learned, too, that in every step there must be a compromise—between what we wanted to do and what we could do within our means and strength. Unlike the city, where we waited for each day to end, longing for the four o'clock whistle, now the days never seemed long enough. Our hearts ached as twilight brought them to a close.

Before, we had been only half alive. Now, lean from hard work and bronzed by the sun, we felt wonderfully limber and strong. Three months had done wonders for us physically. We had spring to our steps, a feeling of well-being. Calluses crossed our palms instead of blisters. That mark of frantic living, the look of tenseness on Jeanne's face, had changed to a deep serenity—she was more beautiful than ever.

Our senses, dulled by civilization, were springing to life. We could tell, by their scent, a dozen different wildflowers and spot a widow-maker high in the arms of a mighty fir that at first had been only a green blur. The simple food had sharpened our taste. It began with Jeanne's homemade bread, which tasted even better than it smelled—not like the flavorless fluff we bought at home. Vegetables picked, cooked, and eaten the same hour taste like nothing from a freezer or can.

But, to top everything, we had a new conception of an egg. Our eggs bore little resemblance to their city cousins. Two beliefs had been shattered: the holiness of a white egg and the food value in a sickly yellow yolk. The color of the shell, we found, had no bearing on the contents—it was correlated with the color of the feathers. But the most astonishing difference was in the yolks. Ours were dark orange and loaded with vitamins from hens that ate natural greens outside the run. Under the old-fashioned grading system, brown shells and dark yolks were both frowned on—a loss of the farmer's pocket and the consumer's health. An egg that comes from an unfettered hen, which you own body and soul, tastes like none you'll ever cart home from a supermarket.

The city had made us immune to sound. At first, the stillness of the forest frightened us. The island's silence made us nervous and apprehensive, conscious that we were a frail bit of matter in an unfriendly environment. The slightest noise startled us.

Slowly, as the weeks slipped by, the silence we thought so great disappeared. The island breathed with constant sound—the rustlings and twitterings in the bush, the drone of bees and insects, and the ceaseless murmur of the sea. We no longer needed the comforting blare of the radio. Life had bestowed upon us an unexpected gift. Our untrained ears had found the magical wave-length that made up our world. From a distance, we could easily identify the marauding seal by its sudden burst to the surface, gasping for air; the weird fluted cry of the loon and the playful barks of the sea otter frolicking in the cove.

By the end of August, our hearing had grown so acute that Kiki roused us from sleep crunching on dry madrona leaves. In Los Angeles, the alarm clock would seldom rouse us. Now, we couldn't stand one in the room.

The veneer of civilization was wearing off. The sun had now become our clock and hunger decided our mealtime. Newspapers and magazines expired. The latest disaster no longer interested us, politics held no meaning, and the phony Hollywood life left us cold. One night we wound up the phonograph to see if we could still dance. With only Kiki for an audience, we made the flimsy floor quake. "It's more fun than the Palladium," Jeanne said when she caught her breath.

Simplicity had stolen into our lives. There was no rushing to the market, the beauty parlor, the laundry, no hours of primping, deciding what to wear or not to wear. The same shorts, hair-do, and shoes did the trick every day. Our cleaning bill? There wasn't any. Nor did we have to worry about neighbors, payroll deductions, haircuts, traffic, income tax, or losing a job.

However, we had fears that rarely touch city people. We never knew when to expect Indians or how long the water would hold out. But our most solid fear was forest fire. The heat made the island dangerously dry. Our only insurance was never to let one get started. We were careful not to smoke in the woods, and we dumped the garbage and trash at sea.

Always present, too, was the possibility that Bertha might not go, and we would be marooned. The island could be as escape-proof as any death cell. So we learned to keep the motor serviced, gassed up, and dry.

Then there was money—the lack of it, that is. Bertha, the trip up, and the current store bill had dwindled our cash to $786.50. It was nearly September, and Wallace had hardly been scratched. But we were too busy to worry. Later, perhaps, when winter closed in, there might be time for worry and quarrels. Now there was only time for work and love.

One night, unable to sleep, we lay awake smelling the sweet scent of honeysuckle and listening to the sea lap the rocks.

"Is there room in a man's heart for more than one great love?" Jeanne said.

"Why do you ask?"

"Oh, sometimes I worry." She put her hand on my chest. "The island has such a grip on you, Dave. It's in your eyes, the way you look at the cove. The way you refer to it as our ship. The way you tackle everything so seriously. I'm afraid—afraid you'll grow to love it someday more than you do me."

The thought was so ridiculous. I laughed and kissed her. "You're jealous, aren't you? Silly girl! No island is a match for you."

The secret of loving is overlooking, something very difficult to do when it comes to crows. Like some people, they are arrogant, crafty, and loud. But in spite of their abrasiveness, they are loyal, courageous, and wise. At daybreak, they swooped over the shack and settled among the trees, the air ringing with their preparations to depart. Sentries like Freddie that remained became our closest friends. A bit after six, he would light on the roof, lean over, and peek in the window. His fiendish squawks routed us from bed with the desire to choke him. Instantly, Kiki would launch an ineffective attack, sneaking up the apple tree to stifle the demon. At noon, we watched Freddie and his gang wrestle scraps from the gulls. It was always a battle of wits. The crows depended on stealth and speed, whooshing down from their madrona blind when an unsuspecting gull contemplated the scene.

Every afternoon around five the commuters returned. Fluttering across the channel like black ribbons, they swarmed into the cove and dipped sharply over the roof to tell us they had clocked in for the night. We loved the crows. They were the guardians of our island home. Even today, they warn us of strangers, drive off the eagles that menace the chickens and newborn lambs, and rid us of starfish, which could easily wipe out the oysters. In the spring, they make short work of the tent caterpillars that threaten the fruit blossoms, knowing in their uncanny way that they will share in a bountiful crop.

One night Jeanne surprised me with a special meal: a clam cocktail as appetizer, filet of cod exquisitely browned in egg batter, peas and carrots, tossed salad, and thick warm slices of homemade bread. Then, from the cupboard, she produced a wild blackberry pie.

"You sure outdid yourself tonight," I said, swallowing the last bite.

"Notice anything different? Besides the fish?"

Suddenly it dawned on me. Everything was from the island! But how had she caught the fish, I wondered.

"Simple—just an old hook, bacon, and grocery twine."

It was a meal we have never forgotten. But, more important, it gave us a sense of achievement and independence that we were to need through the long winter.

Our kingdom was beginning to produce. The garden staggered us with vegetables, and the chickens laid more eggs than we could eat. We were vegetarians and loved it. How grand that feeling of self-sufficiency!

We worked and rushed about with hardly a stitch on, always on the alert for boats. One shining day slid into another. We chased crows from slopes covered with blackberries, to bring home bucketfuls for jam. We picked the first Kings and Winesaps. Jeanne put up applesauce, baked cobblers and pies. Bradshaws and sugar plums filled the larder. The race was on between fruits and vegetables as she feverishly bottled everything she could put her hands on. "We might not have steak this winter," she commented, lifting jars of plums from the pressure cooker, "but we're certainly going to eat." Unfortunately, as it turned out, neither of us realized how long the winter would last.

The room was fearfully hot from the stove and sun, but Jeanne stuck with the task. Her pluck filled me with admiration. I thought how utterly dependent we were on each other, how contented we were, how good it was to live each day being needed. We were tethered to the island and to each other, to love and be loved without asking anything more.

"Let's skip work today and go exploring," Jeanne suggested one morning. "We've earned a little play."

With a picnic lunch, we headed into the forest happy as gulls at low tide. Kiki romped ahead as we jogged along "Park Avenue," our favorite trail, laughing and singing, stopping occasionally to feast on berries and drink in the loveliness of our world. Freddie's castanets clicked among the trees, and the scent of warm fir needles filled our lungs. What a sight we were! Our mouths were blackberry-purple, and our tattered shorts, cut from old pants, did little to hide our nakedness. I hadn't shaved in days.

Jeanne reminded me of Rima, the forest girl in Green Mansions. Her sun-bronzed face radiated a savage vitality, and her long gleaming hair tumbled down her shoulders.

We roamed into the heart of the island, following the scorching meadow and plunging into a jungle of firs. Woodpeckers hammered in harmony to the songs of towhees and larks. A deer bounded into the bush. Farther on, Kiki spied a coon and chased it up a fir. Scarcely believing it real, we came to a fairytale cove like a pool sunk in the forest. From a leaning madrona limb, we splashed into the emerald water. Flounder scooted among the reeds and droves of minnows darted from sight. When we surfaced, a mink slipped out of the water, shook himself, and gave us a nasty stare. With the cries of gulls overhead, we played and explored the underwater jungle, which was as lovely as anything on shore.

"Look, apples," Jeanne shouted, pointing to small red dots hanging from a tree, its scraggly limbs draped with funeral moss. "Beat you!"

I felt my head shoved under. When I came up, gasping, Jeanne was nearly ashore. She rushed to the tree, then stood waving the apples in her hand. We sank onto a moss-covered bank perfumed with wildflowers and munched the forbidden fruit like Adam and Eve.

Suddenly Jeanne shuddered and rose from my arms. Her eyes were deeply disturbed. "I'm frightened, Dave. What if something happens? We couldn't bear to go back to the city again."

CHAPTER 9

The Mysterious Plague

THE NEXT MAIL BROUGHT A LETTER FROM THE CEMENT COMPANY. Eagerly I ripped it open in the twilight.

"Listen to this!" Jeanne leaned over my shoulder. "'Salt is often added to mortar for slower curing, harder cement.' What do you think of that?" I demanded.

"Wonderful. We can start pouring tomorrow."

But shortly after we got home, we noticed something wrong with the floor. I held up the lantern. Milk from Kiki's dish had rolled into the corner. Obviously the floor had begun to sag—probably from the weight of the drum on the roof.

After breakfast the next morning, I crawled under the shack, not realizing what I was getting into. I pried up the beams, shoved under a few rocks and shingles, and came out covered with cobwebs and filth.

"You even smell musty," Jeanne said as she brushed me off. "The tub for you tonight, young man."

The rest of the day we spent building a mixing box, sifting sand from gravel, checking forms, and lugging sacks of cement from the shed. Jeanne complained that it always seemed to take longer to get ready for a job than to do it. By nightfall, when I stepped from the tub, I was so tired I fell into bed unaware of the rash that must have begun to spread over my body.

The next day I would rather erase from memory. It was about noon, and we were mixing concrete when it hit me—a queer, faint feeling. My heart was laboring, my chest burned, and nausea gripped me. A moment before I had felt fine, mixing the mortar briskly, thankful a northerly wind was blowing to keep us cool. Now I was limp and shaking, leaning on the shovel and sweating profusely for no reason.

"Dave, what's the matter?"

"I don't know," I mumbled as she rushed over. "I'm hot and dizzy and itch all over." I tore open my shirt.

"Great Scott! You're red as a lobster. What have you got?" She looked at my legs and groaned. "Looks like poison oak."

"Can't be. There's nothing poisonous on the island."

My legs wobbly, I clung to her arm as she helped me to bed and covered me with blankets. My skin was beet-red now, strangely rough and scaly like an alligator's. I was on fire wherever my body touched the sheets.

Jeanne thumbed frantically through the "home doctor" book. "It's not measles or scarlet fever," she said at length. "I can't find anything to explain it."

Despite my fever, I lay there shivering, stunned at the fact that I was sicker than I had ever been before. But my thoughts were on Jeanne too. How would she manage? Serious illness had never occurred to us. The only medicines we had were aspirin, iodine, and cough syrup. A gust of sea wind fluttered the curtains. The boat! Jeanne had not even run it. Why hadn't I taught her?

The next moment, she was holding a cup to my parched lips, forcing aspirin between them, then wiping my head with a cold cloth. The thermometer rattled in my teeth. Temperature 103 degrees. It didn't seem to matter. I could feel myself fading from the heat as if I were being broiled in an oven.

When my eyes opened again, Jeanne's frightened face was bending over me. Fumes from the gas lantern told me it was night.

"Feel better, darling?" Her hand ran over my forehead. "Not so bad, really." I was lying. "I'll be okay in the morning."

By morning I was in a coma, burning with the fever. I have little recollection of what followed. I rely on Jeanne's diary:

> AUGUST 28: Dave delirious all night. Raved and thrashed so that I had to tie him in bed. I'm really scared. It's too rough to cross for

help, but I wouldn't dare to leave him anyhow.

11:00 A.M.—Tried giving him some broth. His eyes are glassy and his lips all swollen. He's still in a stupor—doesn't even hear me. Every half hour or so he has a severe shaking spell. If only the wind would die down! I've got to try to get him over. So afraid I can't do it.

4:00 P.M.—Dave's not quite so delirious, but he just mumbles and I can't understand a word he says. Bathed his face and gave him more aspirin. Poor dear!

Jeanne's first attempt, late that afternoon, to get me off the island was a complete failure, according to her diary. Even if she had been able to manage getting me out of bed, my rubber legs would not have borne my weight. By seven, the aspirin was gone and my temperature soaring again—to 105 this time. Frightened out of her wits, she tried once more to lift me, and we ended in a crumpled heap on the floor.

As darkness began to descend over the island, my breathing got more and more erratic and I resisted all her efforts to get me to drink water. Jeanne was desperate. Then she thought of the wheelbarrow. Somehow she found the strength to manage the transfer, lumping me in the barrow with my body bent double, arms between my legs, a blanket tucked around my limp, shivering bulk. Getting the barrow down the ramp and me into the boat should have been even further beyond her strength—but she made it.

I vaguely remember a sudden burst of noise and a hazy figure bobbing over me in the twilight. Then a rocking motion, strange and pulsating, lulled me back into a world of nothingness.

When consciousness returned, a strange gray-haired man with a stethoscope was leaning over me, pressing the cold instrument to my chest and slowly shaking his head. He straightened with a perplexed look, and I heard him say, "Frankly, I don't know what it is. But we'll find out. I've asked a skin specialist to come from Victoria—he'll be here this afternoon."

When the doctor left, I felt a hand touch my shoulder. Jeanne was bending over me, her face tear-streaked and pale. "Better?"

I nodded. My eyes were heavy, my mind hazy. I tried to rise but it hurt to move. "What about the island, our stuff?"

"Don't worry, darling. We're lucky you're here. Everything will be all right." My eyes closed and her voice began to fade. "The police boat will check the island. I'll be staying at Vic's."

Later, I awoke fully conscious for the first time. I glanced blearily around. Blinds darkened the little room and a clean antiseptic smell filled the air. I felt drugged, for some of the pain was gone. Then the specialist arrived. After checking my tongue, eyes, and heartbeat, he studied the queer red texture of my skin through a magnifying glass.

"Pityriasis Rosea," he said.

My heart thumped. "Is it serious?"

"It could be, in your case. Your wife tells me you had a wartime kidney injury. It's a rare disease, contracted by people with extremely sensitive skin who expose themselves to old, musty buildings. There's no known treatment. At least it's not contagious, and it lasts for only twenty-one days."

To Dr. Meyer and the staff of the Lady Minto Hospital at Ganges, I owe my eternal gratitude. The doctor prescribed penicillin shots, pills, sulfa salves, lotions, and hot baths. My cowardice was appalling. I cringed every time the nurse poised the needle at my rump, and I dawdled needlessly—to delay the torture—whenever it was bathtime. My body still wept and pained. Only blessed sleep freed me from misery.

The second week, the scabs healed and the soreness diminished, but the itching became unbearable. The only relief I found was lying in the little bathtub at the end of the hall. Now the nurses didn't need to start the water running or pry me out of bed. Instead, they pounded on the door and wailed Britishly, "Mr. Conova', come out of they'a. Caun't you finish *Anthony Adverse* in bed?"

To my surprise, visitors trooped in all day. We had more friends than we thought. How mistaken we two renegade Americans were about these islanders. Hurry robs people of goodness, but these country people had the time to be thoughtful and kind, even to strangers. The Anglican minister came with books and the Word of the Lord; the storekeeper, with candy and magazines; the local constable, with assurance the island would be watched; ladies of the Sunshine Guild, with fruit and cookies; and the dearest of all, with kisses and her comforting presence.

How Jeanne cheered up that sterile, nasty cream-colored room! And how I hungered for her visits, gazing out the window at the yellowing

Our home for the first few days

Vic with his usual grin

Preceding page:
Pristine Conover Cove

Going over after the mail

Cormorants nesting in dead trees along the shore

Jeanne and Kiki at apple harvest time

Dave, Jeanne, and Kiki enjoy a meal of freshly caught oysters

Jeanne and Wallie

Jeanne washing clothes during first days

Jeanne and Wallie at sunset

Baby seal

Jeanne's beachcombing adventures netted her a baby seal. Unfortunately, he lived only a few weeks

Signs at the wharf

Jeanne and David Conover, proud resort owners

Jeanne at the kitchen window with her watering can

Our house

A cozy fire

Cabin during the first harsh winter

Baby sea gulls

Scrubbing the Islander

Our nearest neighbor—the mink that lives in the rocks below the house

Visitors at Conover Cove dock

Jeanne and Davey enjoying a stroll

maples, longing to be outside, and listening for the sound of our car rattling up the hill. Each afternoon she came to sponge my face, rub calamine lotion on my sores, read the mail, and reassure me that I was on the mend, that our future was still bright.

But worry plagued me. I moaned over lost time. Money. The boat. Winter was on our heels, the bathroom not begun yet. I grew restless, irritable, cranky about the food. Toward the end I caught myself, like a little boy, making faces at Nurse Dagleish when she left the room.

On September 20, minus twenty-two pounds and $246, I was released from the hospital. W e returned to the island to find the chickens wild in the trees, the garden a shambles, and the well nearly dry. Kiki's lonely vigil had left her with several mysterious scars.

"Poor thing, I wonder how she got them," Jeanne said. "You know, she's never been too fond of Bennie."

"Could be. But even before I was sick, she was hanging around the chickens as if she was plotting some scheme."

Later, we found that someone else was the plotter, and she was hanging around to spoil the scheme.

By the first week in October, the island wore a cloak of brown and gold; the crickets had stopped singing, and the scent of frost chilled the air. The bathroom plumbing stood in the shed, and we were going over the invoice.

"That doesn't leave us much," Jeanne fumed, snapping the checkbook. "We can't fix up this place and build a cottage too.

"Our comfort can wait," I said, unaware of my mistake. "The first task is to put the island on its feet. When the bathroom is done, we'd better start on the cottage."

We had often visited the site, visualizing a cottage snuggled among the firs like those among the pines at Lake Arrowhead. It would be paneled in knotty pine. The sixteen-by-twenty-two living-sleeping room would have beds on each side of the fireplace, a kitchenette, a Dutch door that opened onto a porch, and two broad picture windows overlooking the cove.

"A log cabin will be more like it. We can't afford a fancy cottage now," Jeanne argued.

"There's no choice. We've got to stick with our plans. Roughing it is out of date—people want all the trimmings nowadays, particularly on holiday."

"You're a bigger dreamer than I thought.... And how will you dream up *that?*" Jeanne waved a finger at the calendar. "Have you forgotten the thousand-dollar payment next October?"

I stood my ground. "A good cottage rental will meet it, not a poor one. We'll manage. We can always borrow." With Wallace for security, we couldn't imagine a bank *not* loaning us money.

"That's not practical, Dave. We're in a tough spot now, in debt already. We'd be going out on a limb."

"We did when we bought the island, didn't we?"

Jeanne nodded and sighed. "At least with a bathroom, we can starve comfortably."

The six-by-eight bathroom began to take shape. The roof went on quickly. Using bought lumber only when necessary, I became a "lean-to" artist—that is, until it came to the hole I whacked out for the door.

"What did you do—mistake a yardstick for a ruler?" Jeanne demanded.

Everything wasn't in the book, I discovered. For example, a saw cut has a significant width to be reckoned with, and it never pays to argue with a square. Because of the sloping roof, the five and a half feet of headroom made us lean, too. Backaches and bumped heads, along with a missing saw, proved to be the price of progress. But the true test of a carpenter, the book said, was "how he hangs a door." That test came the next morning. Though just barely ajar, the old camp door struck the sloping ceiling, and Jeanne was ready to hang me instead.

The plumbing—here I could shine, I thought. Its engineering appealed to my inordinate sense of exactness. Unlike carpentry, where plastic wood and paint cover a multitude of mistakes, plumbing has no place for fudgers. Mistakes not only leak, but the leaks pop up where you least expect them.

Getting water into a building, I learned, is not half as much trouble as getting rid of it. Waste pipes are three times as large. And cutting them with a saw the size of a nail file is child's play compared to threading the monstrous stuff. At night, I crawled into bed feeling I had cranked an airplane propeller all day. The matter of venting—running useless pipes up through the roof—baffled me. The book said it kept the plumbing quiet. Instead, the pipes gurgled, coughed, and laughed at me as the basin water bubbled up into the tub.

Finally, the great moment arrived when I hooked up the most important fixture. Jeanne leaned over me breathlessly. "Think it'll work?"

"Positive," I assured her, tripping the lever. "Something's got to."

A geyser rivaled only by Old Faithful poured onto the new tile floor.

"You used more putty than prudence," Jeanne screamed, running for the mop.

But when we stood appraising the tarpapered room, her face glowed. "Even if the only place we can stand up straight is in the tub, I love it. Now it can rain."

And for days it did. It was as if a great dike in the sky had burst its seams.

"Listen to it come down," Jeanne marveled as we lay in bed several nights later. "If this keeps up, we'll be flooded." Widening pools had engulfed the nearest apple trees and were creeping toward the shack.

"Vic wasn't kidding when he said the rainy season began in October," I reminded her. What he had failed to add was that it lasted all winter. Already the chilly dampness had crept through the cracks and made getting up a Spartan affair. We changed into long underwear, rubber boots, and mackintoshes, and Kiki proudly displayed a new furry coat.

The maples stood bare and bleak, portents of advancing winter. Even the cove had changed. It looked twice as large and not nearly so friendly. High tides swallowed the beaches, and logs rumbled ominously against our bluff. Overnight, fall's bright colors faded into a misty world of drip, drizzle, and somber grayness. As our vision narrowed, the audible world took its place: the drumming on the roof, the thundering seas on the points, the whining wind that seeped through the cracks, and the splatter of water from the gutterless eaves.

Then what we were most afraid of, happened. One morning we tumbled out of bed into ankle-deep water. Our shoes, suitcases, and belongings were awash, Kiki sat morosely on the stove, unmoved by the whole affair, while we flew to the window.

The yard was a lake afloat with sawhorses, two-by-four's, and fuel cans.

"Come on," I said, struggling into my clothes. "We haven't much time before this cracker box floats off its foundations." We waded over to the bank that was blocking the water and hurriedly began digging a trench.

Our troubles were not without lighter moments. The sudden burst of water from the trench into the sea swept Jeanne off her feet, and she came sailing toward me, arms and legs thrashing, something like a hooked fish. Barely able to keep a straight face, I grabbed her and hoisted her upright. "Isn't it a bit chilly for a dip, lady?"

By late afternoon, the water rushed down the slopes behind us into our network of ditches and streaked the bay with mud and leaves.

"How will we get things up from the wharf?" Jeanne worried. "The yard's a mudhole."

It meant a lot of work, but there was only one solution—a low bridge. I made a sketch that evening. "We'll bring up boulders and anchor them in the mud." I explained. "Then, with a few timbers, lay down planks for a road." It had to be wide enough for the cart that I had made from the wheels of Vic's old Essex.

For days, we collected drift and barrowed rocks up from the beach. With beams from an old pier, we wallowed in the mud and spanned the swamp, the rain constantly beating our faces, our hands so bruised and grimy we could hardly spike the planks.

The first day of November brought a truce between island and sky. The roof was silent. The morning broke bright and brittle with frost that reddened the cheeks and made the ground gleam and crunch underfoot. A thin gauze of smoke hung over the cove, scenting the air with the sweet smell of burning wood.

Kiki was watching me chop kindling when Jeanne came out, shivering and clutching her apron.

"Pipes are frozen. Would you get me some water?" Her face was drawn, and I could tell she had been crying. "I don't know why this island is such a headache. Always something bugging you. Nothing but work—work—work! I'm *sick* of scrubbing." Her voice grew bitter as she held out her hands. "Look at them! They'll never be the same. Do you hear me? *Never!*"

Stunned, I reached out to grasp them. She buried her head in her apron and fled inside, sobbing. By the time I returned from the well, she had pulled herself together.

"I'm sorry," she said, trying to smile. "Sometimes things get the best of me and I have to let go. Everywhere I look I see work, a grim future. It's too much, Dave."

I swallowed her in my arms. "Darling, I know. It seems as if anything worthwhile demands more sacrifices than we can bear. But we'll make it, believe me. And when we do"—I lifted her chin—"it'll be worth all the headaches."

Two weeks later, when the frame of the cottage was nearly up, work came to a halt. The shed was bare. After dinner I tapped the barometer.

"Still high and rising," I said. "Better get your list ready—I'm going in to Victoria tomorrow. Think you'll be all right overnight? "

"I'd feel better if I didn't have to lug that rifle around, but I'm afraid to go anywhere without it."

"You need a pistol."

"There's too many other things we need. Skip it."

But I didn't. At daybreak when we untied Bertha's icy ropes, I had the Contax hidden in my pocket, with a trade in mind. I could just make the eight o'clock ferry.

"Be careful now," I warned. "And leave off the beachcombing."

Jeanne promised meekly.

CHAPTER 10

Lesson from the Sea

VICTORIA WAS NOT THE GAY TOURISTY TOWN of tallyhos and flower baskets that it was in summer. A dismal rain had rubbed off its Old World charm. The narrow streets and age-old storefronts were dreary and oppressive, and people with their heads bent trudged the sidewalks like pallbearers. Even the air had an unpleasant oily smell.

On Douglas Street a new office building was going up. Wartime shortages, I thought, couldn't be as bad as Vic pictured. With a list of addresses from the telephone directory, I started on the rounds of the lumberyards.

"Sorry, we can't help you. We can't fill orders for our regular customers. Afraid you'll have to try elsewhere," sober clerks repeated. "Sorry, we haven't got a nail."

What a bitter disappointment! I hated the thought of going home empty-handed. Jeanne was counting on me.

Night fell before I half finished my list. In a dreary hotel room I fell exhausted into bed. But I couldn't sleep, thinking of Jeanne on Wallace alone, without a boat. Was she all right? What would she say? I smoked myself hoarse, tossed and turned to a myriad of street noises and the clatter of English cars, trying to believe tomorrow would be different and wondering if the odds were not piling up against us.

The next day I began to loathe Victoria. The same answers greeted me again. By the time I came to the last firm on the crumpled list, I was badly dejected. Why bother? I had only an hour to make the last ferry. Still, I felt an urge to try it.

The Drysdale Builders Supply Company was an old Tudor house, plainly a family affair. I entered the door still gripping the soiled list in my hand. The empty office reeked of cigars; from the ceiling dangled a feeble light bulb. Suddenly a balding, Buddha-shaped man with a stern face barged into the room. His rolled-up sleeves and rumpled white shirt told me he was not just another hard-hearted business man.

"Sorry to keep you waiting," he apologized. "Had to take a rush order to our mill in back. I'm Lester Drysdale. What can I do for you?"

I told him my difficulties. My depressed state must have moved him, for he ushered me into his private office, offered me a cigar, and pointed to a chair.

"Where are you from?"

I poured out my story while he shifted his cigar back and forth in his mouth, and his steel-blue eyes grew larger.

"You Americans amaze me." His look was a mixture of wonder and disbelief. "That's a pretty big job for two people. Aren't you tackling more than you can handle?"

"Not if we can get everything we need."

His shrewd eyes studied me. "Give me your list. Maybe I can help you." The building materials we needed filled two sheets of paper. He scanned them carefully. "Some of these things are expensive and hard to get. What would you say if I came over Sunday and looked at your layout? Maybe I can save you some money."

I sprang to my feet. "Would you? Golly, I'd appreciate that."

Jeanne's diary for November 2, the day I returned, read:

> Last night was pure torture. I quaked every time an owl hooted. Kept the lantern on and took the rifle to bed, but didn't get a wink of sleep anyhow. Today I lugged the gun everywhere, afraid an Indian might jump out at me. Kiki sticks to me like glue. Water stopped at noon, and it took me an hour to prime the siphon at the well. Whitewashed the chicken house till dusk. Too nervous to stay outside any longer.
>
> It's nearly dark now, and I've got the lantern lighted—my hands

shook so I could hardly manage it. Have started dinner—but where *is* Dave? He said he'd be back before dark. Why doesn't he come?

It was eight o'clock before the drone of Bertha's motor finally signaled my return. By then, the fire was out, the dinner ruined, and Jeanne had worried herself into a panic. Grabbing the lantern, she rushed pell-mell to the float and flung herself into my arms.

"Darling, I'm so thankful you're back—you're safe. Don't ever, ever leave me that long again."

For a small eternity we clung together while Kiki purred joyously against our legs.

The next night at dinner I put a parcel beside Jeanne's plate. It was her birthday. Eyes dancing, she pulled off the wrappings.

"An automatic! Oh Dave, you shouldn't have. How did you—"

"That's not important," I cut in. "You needn't be afraid now."

She leaned over and kissed me tenderly. "I'll start practicing tomorrow."

In no time, Jeanne became a regular Annie Oakley, shooting cans in the air and bottles at a hundred yards. With the cowhide holster on her belt, a snapping twig no longer startled her and she ventured farther afield alone. I was glad, though she did give me one scare, and I spoke to her about it: "Unless you want to become a widow, madam, you'll have to give up practicing in the cove. Your shots ricocheted off the rocks this morning and just missed my head."

On Sunday we met Lester at Fernwood. Sporting a Stetson and his usual fat cigar, he looked like a Texas Senator out to inspect a government project. The morning was made to order—the sea and sky contrasting hues of gentian blue. We hoped he would rave about the island. However, he made no comment when we rounded into the cove, nor even when we docked. Only amazement showed on his sober face. Right there, we began to understand the big difference between us and Canadians. They often leave things unsaid that speak for themselves.

"Had no idea the island was so large," he said finally, when we were walking up the trail. "You've got something here, Dave."

As we showed him around and explained our plans, he penciled rough sketches of cottages, store, and recreation lounge. Impaled in a cloud of cigar smoke, he bounced from one location to another, pointing out the best exposures, explaining the grades of lumber, and recommending log siding, Gyproc, and other materials a beginner could easily handle. By noon, we were too excited to worry about money.

Over tea and sandwiches Lester eyed the room. "To be frank, I don't know how you kids will do it. Judging by your quarters, you haven't much capital—but I admire your courage," he said with a smile.

I told him that we'd have to manage as best we could, and he promised the order would be delivered in a couple of weeks. When we parted at Fernwood, I was sure he wished he were staying instead of saying good-bye. We felt that we had found a new friend.

"I hate to think of you spending days lugging that stuff over," Jeanne worried that night. "The weather's too unpredictable. It's too risky. Isn't there some easier way?"

"We need a small barge," I answered, dreaming.

"What about a raft? That's almost the same thing—we could build it ourselves."

"And we could use it for hauling sand and gravel too," I added. "Why didn't we think of it before?" However, it was always in solving the simplest problems that we committed the worst blunders.

In the morning we started looking for cedars that would make suitable float logs. Cedars, we discovered, did not grow alongshore where they could be easily felled into the water. They throve in small valleys and low spots a bit inland. After a day's search, we located three sisters in a draw not far from the sea. Next morning, carrying axes and a peavy, we stood under them contemplating their fate.

"They're so graceful it's a shame to cut them," Jeanne said. Then she turned with a puzzled look. "How'll we get them in the water?"

The draw was choked with waist-high salal, willows, and vines. I could tell by the husky look and feel of a peavy that it was a logger's best friend, but I had no idea whether it and a few rollers would do the job.

"We'll worry about that," I told her, "when they're down."

My ax bit into the first tree. Chips plunked into the salal and the sharp pungence of cedar stung my nostrils. As the notch deepened, sweat rolled into my eyes. There was a groan, a slight movement, then a splintering sound.

"Timber!" I yelled.

Jeanne looked up at the quivering branches. "Which way is it going?"

"Lord knows! I'm no prophet," I yelled, grabbing her arm, and we dove under an outcrop of rock. The tree crashed the wrong way up the draw.

While I tackled the other two, Jeanne lopped branches off the first one and cleared a path to the sea. By late afternoon, with three logs twenty feet long and over a foot thick lying at our feet, we began to consider the problem of moving them. Instead of trying to roll them up and over the rocky ridge into the sea, we decided to push them, on skids and rollers, downhill to the little beach about one hundred and fifty feet away. Jeanne pried with a fir pole while I pushed with the peavy. For all our heaving and groaning, we could gain only a few feet before we were out of breath and had to rest. The ground was soggy and uneven. The logs got mired, and we had to pull them out with block and tackle. It was three days of grueling toil later when we succeeded in rolling the last one onto the beach. Tired, dirty, and badly bruised, we collapsed beside them.

I handed Jeanne the water jug. "Still interested in a log cabin?"

"Are you crazy?" And we both began to laugh.

I studied the tiny shell beach flanked by giant firs. We could make the raft there, I decided. With rollers and a high tide, we shouldn't have any trouble getting it into the water.

When the logs were lined up together like strips of bacon in a frying pan, we began scouring the shoreline for lumber. After a day's search, we had collected enough boards to go to work.

On the seventh day, the twenty-by-eight-foot raft was completed and waiting for high tide at 6:20 P.M. Shortly after the sun set, we brought Bertha around to help. We could barely contain ourselves. At 6:15 the sea began to nibble at the logs. With Jeanne at the tiller, Bertha roared, tightening the rope, as I pried from the shore with the peavy. Squeaking and groaning, the raft slid into the water.

What happened next still makes us blush. It sank below the surface like a diving whale. I could not believe my eyes. The next instant, it bobbed up—its decks awash. Waving to Jeanne to cut the motor, I raced out and climbed aboard the raft. It listed precariously and began to go under again. I had to balance myself in the middle to keep from sliding off.

Jeanne pulled alongside. "What's wrong with it?"

"Damned thing has no buoyancy. The logs are too small."

They were also too green and heavy. Later, we learned it would have helped if they had been dried out for a year or we had cut larger trees. Why hadn't I given more thought to this project? Study and planning, I knew now, were the keys to this life. Broken-hearted, we beached the monster in the twilight; it remains there today—a monument to lack of foresight, useless except as a lesson.

I sent away for more library books, and at night pored over their pages until I couldn't stay awake. Full mastery of the job was my goal.

"We've made so many blunders," Jeanne said one evening at dinner. "We just aren't cut out for this life. Don't you ever get discouraged?"

I looked up. Her smile failed to hide the worry in her eyes.

"What about the biggest problem?" she went on.

"Money?"

"Yes, money." Her voice rose. "We can't go on much longer. What if we can't get a loan?"

I sensed Vic's words were on her mind. "That's not likely. Vic's been wrong before. Don't worry"—I patted her arm—"banks will lend us money."

In the morning we were startled by the frantic squawking of the chickens. Jeanne grabbed her gun and we dashed up the trail. Four hens lay dead with their throats slashed, and the survivors were running about the yard clucking in terror.

"Mink!" I exclaimed. "The dirty rascals!"

There was rustling in the bushes, then a savage scream. Jeanne aimed her gun.

"Hold it!" I ordered, creeping closer.

All was quiet for a moment. Then the salal leaves parted and Kiki hobbled out, her head gashed and bleeding. She began to purr as if she knew she had done something brave.

"Poor dear." Jeanne scooped her up in her arms. "She got the worst of it, Dave. Too bad that murderer got away."

I was brushing aside the salal. "I wouldn't be too sure of that." I stopped. "Looks like you've got that fur collar you wanted."

Kiki writhed in pain while we washed and sewed her wounds. The loss of blood made her pitifully weak. For several days it was touch and go. She didn't budge from the box under the stove. We were worried. The island would not be the same without her. But one morning she limped to the door and meowed. A diet of clams and oysters had restored her.

The next day the building materials were at Fernwood. The mound of lumber, plywood, shingles, and bricks nearly took our breath away.

"We didn't order all that," Jeanne said.

I scanned the pile. "Must be some mistake. These screens don't belong to us. And I don't remember so many kegs of nails."

The mail brought a letter of explanation:

Dave,

Thought you could use a few extras. Don't worry—I've marked them lost in shipment. Get busy. I haven't had a good holiday in years.

Yours,
Lester

P.S. The screens will make good windows.

It was good to know someone was rooting for us. However, our spirits soon ebbed. The Fernwood wharf was growing dangerous. As we loaded the boat, the cart wheels kept sinking deeper into the rotten planks. "We'd better put down some drift boards on top," I said, "or we'll be minus a lot of stuff."

"Think they'll ever get around to repair it?"

"Eventually—things move slower here."

When the boat was full, we threw Roberta's canvas over the lumber pile. The sun had just slid behind the trees.

"Think it's safe to leave all this?" Jeanne asked.

"According to Vic, there's nothing to worry about."

"Speak of the devil, here he comes. We'll ask him about the wharf."

Vic strutted up like a mechanical toy, thumbing his braces and casting a scornful eye at the pile. His blue eyes twinkled as his foot poked the lumber. "Puttin' up a henhouse, Tiny? Them boards got more knots than wood. Ain't fit fer a barn."

I smiled. "It's knotty pine, Vic. More knots the better."

A shot of tobacco juice hit the ground. "Yuh plum out yer head? All my life I've been avoiding knots. Now yuh want me to believe them things are good. Gotta be careful, son. That island is makin' yuh bushy."

Jeanne and I both grinned. Then we mentioned the wharf.

"Ain't likely they'll fix it. Nobody much usin' it anymore."

The laborious ferrying began. Everything had to be carted down the trembling four-hundred-foot pier and packed down the gangway into the boat. The other side was even grimmer. On numerous occasions our little float sank, and hauling twenty-foot fir six-by-sixes up the steep ramp was a back-busting task. In all, every board and brick had to be picked up six times. But there were compensations. On Wallace, Jeanne was always on hand to help. With a harness fashioned from an old inner tube, I pulled the loaded cart, donkey-style, up the trail while she pushed the rear. Grimy and panting, I was always a weary, bedraggled sight when we reached the shed.

"Wait, I'll get the camera," Jeanne said, laughing, after one trip. "If Bill Harvey could only see you now!"

"You needn't bother," I mumbled thankfully. "We haven't got one."

There were times when I failed to load Bertha properly. Too much lumber on the bow lifted her stern. Her prop thrashing, she wallowed helplessly in the choppy sea like a wounded goose.

Then I committed a prize blunder. Coming across on a calm day with a mighty load, I was deep in thought—not watching where I was going. Wham! Bertha smacked into a drifting tree. What a fool I must have looked, sitting there engulfed in its branches like a fly in a spider web.

For me, there was deep satisfaction in being on the water. I loved to lie on the lumber, my face into the wind, and feel Bertha plow into the chop while I listened to the sea chuckle against the planks and watch the green water glide by. My senses were musical instruments on which the sea played her symphony. Some days were so serene, so still and shining, that they shook your faith in Kodachrome. These sea-borne moments of gleam and glitter and sparkling blue were like another life, a world in themselves. Freedom was a tangible thing. The mind could travel out and up, unimpeded, free as a gull.

Late in November we brought Myrt and Vic over one beautiful day and had a picnic lunch on the beach. Vic was in his element. He strutted about, thumbing his braces and telling me how I *should* do things. By evening, my pride was distinctly wilted.

On November 29, Jeanne noted in her diary:

> My mail-order woolies came today, but they'll have to go back. When I tried them on, Dave told me I looked like an overstuffed Christmas stocking. Of course catalog shopping is always a gamble—here, at any rate. You never know what you'll get. Order size 28, you get 38. Order red, you get blue. And you also never know whether an item will be English or American or—surprise!—Canadian. It really sets Dave off. He has American plumbing books, then gets an English Worthmore toilet with instructions in French. What a country!
>
> Wish I knew what to fix to eat. Whenever I ask Dave, he always says, "Whatever you feel like, honey." He has strong views on everything except what he puts into his stomach. Yet if dinner is ten minutes late, I get the quiet treatment right through dessert. When the food grabs hold, he's more like himself—charming, polite, and *pigheaded.*

By now, most of the supplies had been ferried over. With the end in sight, I got a bit too cocky. Late one afternoon I decided to take all that remained. The gunnels were barely six inches from the water when I pulled away from the wharf. I was sprawled on the long lengths of lumber over the motor, my hand clutching the tiller cable. The clear sky and light southerly ripple gave no hint of trouble.

Halfway across, the wind shifted and began to freshen. Soon it had the earmarks of a northerly squall—the hardest to predict and the wind I feared most of all. There are no low, black, scurrying clouds; no gusts or howls, nor marching rollers that warn of a southerly gale. The northerly strikes suddenly, with no warning, when the sky is bluest and the air sharpest. It has an eerie, frightening quality; it's rarely heard, only felt. The wind and sea are one, the air liquid, the visibility nil. The waves, like the prongs of a pitchfork, are close, sharp, and steep, not the mountainous snarling seas of a southeaster. In a small boat, you go through them instead of over them; you are half in, half on, the sea—an experience you'll never forget.

The waves grew steeper now, bursting in white arcs over the bow, blurring my sight and soaking me through. Bertha shuddered deeply, her exhaust grumbling. Prostrate on the lumber and cold as death, I held onto the cargo rope and wondered how long she could take it. She was getting groggy—I could feel it. Her bow plunged deeper, each icy deluge threatening my grasp. I squinted at Wallace through a rent in the rushing green wall, and blinked in disbelief. Had the island broken loose from its moorings? It was getting farther away. When I glanced quickly back at Salt Spring, my heart sank. I was going backward full-speed-ahead. There was only one thing to do—a chance in a million: Keep her bow into the waves and try to drift back toward the wharf. To attempt to turn would be suicide. Everything depended on a firm hand.

With no leeway now, Bertha's nose swung perilously one way, then another. Her buoyancy nearly gone, she breasted the waves with a groan, tearing them apart like a surfacing submarine. Dropping backward blindly was by guess and by God. Every minute seemed an eternity as visions of reefs and rocky shores rushed to mind. What if I missed the wharf? The air was fluid with sheets of spray. Salt blinded my eyes and choked me. All I could see was tumbling white water, water that surged over the bow, slashed my face, and froze me to the lumber. Desperate and scared, I struggled to my knees, my jacket ballooning like a sail.

Then relief swept over me. Fernwood was perhaps a block-length away now, the wharf bouncing like a plastic toy as the sea broke over it. But it was impossible to dock alone. My eyes raked the Bettiss farm. Were they watching? I hoped to God they were. Any moment the motor might stop or an angry comber end it all. Bertha neared the dancing float, and again I frantically searched the house, the outbuildings, the meadow where Vic's cows munched peacefully.

At last I saw it. A dark figure was running down the road to the wharf, waving its arms. Above the wind, I heard Vic's yell: "Wait a minute, Tiny!" The next instant he was on the bobbing float, bracing his feet against the float rail and grabbing the bowline I flung into the wind.

Together we secured Bertha to the lee of the dock where the wind would hold her off so that she wouldn't bump.

"That's cuttin' er pretty close, son," Vic scolded, wiping the hair from his eyes.

"You're not kidding." I pumped his hand in thanks. "You're right. Bertha could use a lot more go." But until we could afford it, I knew the motor would have to do.

As if to mock me, the wind stopped as suddenly as it had come.

The sea swiftly changes a lifetime of confirmed beliefs. The straightest line between two points is never the shortest, rarely the quickest, and seldom the safest. In an open boat, you become a quick learner or a lost pupil. For the sea is the only school that never gives you a chance to fail a grade.

CHAPTER 11

Mail Order Christmas

THE FIRST WEEK OF DECEMBER BROUGHT A FORETASTE OF WINTER. We expected cold-days of freezing temperatures and a little snow—but not weather that attacked and numbed the spirit more than it touched the body. Lord, how it rained! And rained! We thought the world had gone off its axis and we were doomed to a sunless life. The forest around us was like a wet sponge; water oozed, trickled, and gushed into the cove from every nook and gully.

"No wonder everything grows so large," Jeanne said.

"It's the price of beauty," I pointed out.

"You can afford to be philosophical. Wet wash and dirty floors don't bother you." Her lips tightened.

"You can get used to—"

"Skip it. That doesn't work for everyone."

For days the island lay shrouded in perpetual dusk, days so dark and drizzly it was difficult to tell when the nights began. Our world shrank to the radius of the gas lantern; from the window, Salt Spring looked miles away in the silver haze. When the sun's timid rays crept out, it was like taking off very dark glasses. Colors leaped out in the brittle air: the flesh-pink madronas across the cove, the emerald lichen on the rocks; even single firs on Salt Spring stood out. Then, almost in moments, the cove would be blurred by a drizzle of rain and the wash had to be hung back over the stove.

Rain drove us inside and slowed progress to a crawl. The windows wept, puddles formed under the door, and the damp crept into food and clothes. Everything smelled dank. Bessie ate more wood; the lamps, more gas, and Bertha always needed bailing. To stop the invasion through the cracks, we papered the walls with layer upon layer of newspaper. For contrast, we put the Sunday comics on the ceiling with Lil' Abner over the bed. The effect was colorful, but did little to warm our backs.

We hovered about the coffeepot, blaming all our troubles on the weather. "If it would only stop raining," I harped, "we could get on with the cottage."

And Jeanne would add, "I could hang out the wash or catch a cod."

The weather controlled our lives. It decided what we wore, what we did, and—often—what we ate; even how long we could stay in the tub. On a few occasions, it graciously permitted us to dig in the garden. More often, it sent us rushing to cover the tools in the shed. Sometimes it kept us up half the night watching the boat. It also dictated whether we were prisoners or could go to town.

The barometer was consulted a dozen times a day. "It's a wonder the thing works at all, the way we bang it to death," Jeanne said.

Lost time began to bother me. The rain hadn't stopped for days, and I had read all there was about hips, rafters, and miters. Throwing on my mackintosh, I started for the door with the announcement that I was going to work.

"You're crazy! It's coming down in buckets. You'll break your neck on that fool roof."

"Doesn't matter," I replied. "It won't go on by itself."

In spite of the downpour, I began cutting rafters. The saw balked in the wet cuts as the drizzle ran down my neck. The only way to live in this country, I muttered to myself, was to ignore the rain. We were just spoiled Californians. The natives did. Why shouldn't we?

It had puzzled us why the natives looked so sober and found it difficult to smile. We blamed it on British stuffiness. Now we knew it wasn't that. When you don't see the sun day after day, sometimes week after week, it's bound to affect your disposition. To make you turn inward, wrap yourself almost unconsciously in a hard-to-penetrate protective shell.

Jeanne and I felt it happening to ourselves. One night shortly after we had gone to bed, she mentioned it.

"Dave—don't you sometimes think the island is too much for us?"

"Heavens, no. Why?"

"Oh, just how we've changed. We don't talk as much. I never know what you're thinking anymore, and you never want to come to bed. It makes me depressed and lonely—as if the island is driving a wedge between us."

I knew she spoke the truth. A woman can mold to a man, though never completely to his life. But I couldn't let her get discouraged.

"It's this cursed weather, darling," I said. "Try to ignore it. Lord knows, it's not easy, with all this wet and gloom. You'll feel better tomorrow when the sun comes out."

"I hope you're right, Dave." And her voice trailed away.

For days, storms rocked the island and churned the sea into an inhospitable moat. Not even Bertha was safe in a gale. We had to brave falling branches to fend off drift logs that battered her sides. Around us, six-story firs whipped and groaned like dark pendulums against the sky. The nights were always worse. Snapping limbs brought us bolt upright in bed. Was Bertha safe? The canvas over the cement? Close and warm inside, we could feel the shack quiver like a ship in a storm, hear the wind wailing against the walls and rattling the door.

On December 15, Jeanne wrote in her diary:

> Winter here isn't what I expected. It's either raining or blowing, often both. I'd pictured clear, sharp days outside and frosty nights. Instead, I'm confined to four walls, a foul-smelling lantern, drooling windows, and muddy floors. Dave's so preoccupied it's like talking with a deaf mute. He's never lonely, never tired or depressed. Even in ditchdigging, which I *know* he dislikes, he can find some good. "It gives me time to think," he says. Only time I know he's still the man I married is when he hops into bed. When a man's not interested in sex, you know he's sick.
>
> How Dave gets by on four or five hours' sleep is beyond me. He just plods on methodically, tired or not. He's even worked out a system to beat the weather—lists for inside and outside work, my work, his work, *our* work. He sinks his teeth into everything like a bulldog.

> The chickens have gone cannibal and we have to pounce on every egg. I warned them that if they didn't behave, they'd end up in the pot. Tonight Isobel did. And that reminds me—I took inventory today. All we have left is half a sack of potatoes, three boxes of apples, six jars of peas, and four of pears. Depressing. We should have planted a larger garden, but I don't see how I could have managed more with all the other work. That's the trouble—everything here is work, work, work!

Every night the lantern burned late. I studied to be fit for the morrow's work, and to utilize every moment I blocked out the next day's agenda. How wonderful to dole out time instead of being a slave to it! The chores came first: the woodpile, boat, lanterns, stuck drawers, and broken tools. The greatest bugaboo—it still is today—was estimating the time to do a job. Hauling materials took nearly as much time as building. The bathroom, for which I had figured three weeks, took almost two months. Even doubled, my calculations were often faulty. For an island is always working against you, forever trying to escape by foiling your plans.

During the long winter nights we learned much about ourselves. I had never been a student. Now I lived on books, finding knowledge and know-how a thirst I could not quench. The basic necessities—food, shelter, and water—took on new and exciting interest. Each day became a voyage of discovery. And with each new field I studied, I learned more about myself. Ditch-digging exposed me to geology, which later led us to water—but not without a bitter lesson in resourcefulness.

I marveled at Jeanne's extraordinary dual nature: a tomboy by day, a woman at night. The transformation occurred at dinner. The tousled-headed, smudge-faced lad in dungarees and gray sweatshirt who passed up shingles all day was gone. Across the table sat a lovely creature in a plaid skirt and fresh green blouse, her face glowing in the candlelight like a bright flower.

"Dave, you're pushing yourself too hard," she said one night at dinner. "You come in so tired and beaten. Then all you want to do is study. Play is as important as work, you know."

This was something my puritanical background would never let me understand. "Hard work never hurt anyone," I argued.

But Jeanne was right. The days and nights were never long enough for me. I hated giving up, even to go to bed. When her calls failed to budge me from my studies, her nightgown—one of these sheer transparent things—usually did.

At times like these, I wondered why she had married a shy, introverted, bull-headed soul like me, whom the good Lord had overlooked when he passed brains around or the ability to annex money. And I thought how difficult it must be for her to live with a man so methodical she could guess his every move; a man who invariably rose at the same hour, insisted on porridge and toast every morning, who even wore the same pants day after day (though cleaner ones hung ready on a hook).

Jeanne's spirits ebbed and flowed like the tides. On sunny days, she was a warehouse of ideas, chatter, and bounce. When things didn't go so well or the wet and wind kept us inside, there were long periods of silence and disillusionment in her eyes.

Late one night she put her arms around my neck. "Think it'll be done by June?" she asked, glancing at the cottage plan on the desk.

I nodded, deep in fireplace construction.

"Wish you'd take as much interest in money as you do in books. Last month's bills are still in your drawer. You haven't even looked at them. Do you know how much we have left?"

"Don't like to think about it," I mumbled.

"That's just it! You don't think about it enough. You let me do all the worrying. We've got $84.60. How long do you think that'll last?"

"Don't worry, darling"—I turned to her, slightly annoyed—"a loan will take care of everything. Soon as the roof is on, we'll see about it."

Unfortunately, things didn't work out that way. A week later when I went in to Victoria, every bank manager wanted to know what security we had. An island? They shook their heads. "We don't lend money on real estate." Loan companies asked if I had a job. I wanted to say, "Yes, a big one." Instead, I turned and walked down the stairs.

Lester was more sympathetic. "Wish I could help you, Dave, but every cent I've got is tied up in the mill." He pulled out a twenty-dollar bill. "Would this help?" I shook my head.

Poverty was only something we had read about in *The Grapes of Wrath*. We had never been broke or hungry. But now the full impact of the situation hit me. I was frightened and mad-mad at the Canadians for their stupid banking laws, their stuffy conservatism; frightened at the shadow that had fallen over our dream. How, I wondered, would Jeanne take it. We had to go on. Between the island and the sea, we wouldn't starve. I was sure we could get money *somewhere*. After all, didn't I believe that dreams materialize for those who deserve them?—who fight for them?

When I pulled into the cove, Jeanne was on the float. "Did you get the loan?"

My face gave me away.

"You know what this means, don't you?" She bit her lip to keep back the tears.

I nodded dumbly.

"And I was counting on such a wonderful Christmas."

It was true. She had desperately wanted the first one to be special, one always to remember.

It was difficult to comfort her that evening. There was a lost-child look on her lovely face, and though I buried my discouragement in study, she was unable to settle at anything. The sewing needle kept pricking her finger, and when the silver she brought out didn't need polishing, she threw herself on my lap.

"What'll we *do?*" she cried, burying her head in my shoulder. "We can't go on."

"Of course we can." My hand felt heavy as I stroked her hair. "It's not that black, darling. The cottage will bring in seventy-five dollars a week. Then there's the boat we can rent. And fishing parties. If we can hang on till June, we'll have it made. It'll be rough, I know, but we can do it."

"What about furnishings . . . food . . . gasoline?"

For a second I groped for words. "Didn't you know island owners are supposed to be wealthy? We've got lots of credit." I took her head in my hands and kissed her gently. "Don't look so sad—you're not near so pretty when you cry."

A thin smile parted her lips.

"That's better. Now let's go shopping in the catalog and make Christmas a real fling. What do you say?"

Her face brightened.

We were amazed at how much we needed. The order blank was soon filled with gloves, underwear, and curtain material. Rubber boots never looked so exciting, and cheap cotton frocks glowed with Fifth Avenue appeal.

Jeanne picked out an old-fashioned oil lamp. "It's beautiful compared to our smelly lantern, and it says no flicker or fumes."

"For ninety-eight cents"—I pointed to a cigarette machine—"we can make our own." Jeanne had laughed at my clumsy attempts to roll one. Only necessity would make us give up the filthy habit.

I watched as she totaled the column. "That about shoots the works," she sighed. "I hope we aren't crazy."

"It shouldn't take longer than a week to come from Vancouver," I said, forgetting that, on an island, nothing goes according to your plans.

After a week we began to worry. In the corner sat a Christmas tree decorated with popcorn garlands and tinsel from cigarette and chewing gum wrappers. Cedar boughs and alder cones colored with red nail polish hung from the door and windows. We haunted the mailbox and asked at the post office. Still no parcel.

One day when I returned from town, Jeanne was not around. I called, and her answering shouts led me to the bay across the island. I found her between the sea-washed boulders, clutching her leg, which was caught in the rocks. The incoming tide was nearly at her side.

"Oh, I thought you'd *never* come, Dave. Hurry—get me out of here. I'm afraid my leg is broken."

When the rocks were pried apart, I lifted her to higher ground and examined her leg. Fortunately, it was not broken, but it was badly bruised and bleeding profusely from several deep cuts. While I applied a tourniquet with my handkerchief, I asked how it had happened.

She pointed to a broken board that was not awash, and said shakily, "I didn't know there was a hole underneath."

When the bleeding stopped, I carried her to the shack and got out the first-aid kit.

The day before Christmas, I battled an angry channel and begged the Ganges postmaster to look again.

He did, and reported, "There's nothing here for Conover or Wallace Island. You sure it's comin' by mail?"

"What other way can it come?"

"By boat." He nodded toward the wharf. "Try the freight shed."

The door was wide open. There, on the straw-covered floor amid bundles of stove-pipe and a crated bathtub, was our package. The place was deserted, and the sliding door, off its tracks, leaned wearily against the wall.

My joy suddenly turned to wrath. What inefficiency! Wait until Jeanne hears about this, I fumed. Then it struck me—the incredible honesty of these islanders. Here was one place in the world there was no need for locked doors. When goods arrived on the CPR boat—whether it was a pair of slippers or a mahogany dinette set—the owner could pick them up in a week, a month, or a year, knowing they'd still be there whenever he chose to come. L.A. was never like this. No wonder Vic left the key in his door, that roadside stands were unattended, and that the only constable spent most of his time fishing. Honesty was like the weather here—something we had to get used to.

Yet, Ganges was something else again. The villagers just wouldn't warm up. When I walked in, a hush fell over the store. Wading through whispers and icy stares, I would flash my rosiest smile, only to be asked, "Haven't you gone yet?" or "Won't you be heading south soon?" The man in the lumberyard was more emphatic. Rushing over in a Mackinaw that nearly hid him from view, he demanded, "You still here, Mr. Conover?" It seemed my presence was growing more objectionable with every visit.

"You'd think by now," I told Jeanne, "they'd thaw out a little. It's as if I carried the plague."

She thought it over. "If you ask me, that old scalawag has something to do with it. Anyway, don't let it bother you." She began to grin. "Remember what you keep telling me?"

Christmas morning we awoke to find the pipes hopelessly frozen and the firs bright with frost. I had to pack water from the well, but trouble never seemed so bad when the sun was on my back. I considered how we could celebrate the day. A boat trip? Explore for Chivers' gold? No. We could try to catch our first salmon.

An hour later, we were on the water with the pole over the side. The sharp air tingled our faces as Bertha throbbed along the shore.

"Think anything will bite that shiny metal?" Jeanne looked doubtful. The Deacon had assured me it was the most popular spoon.

I nodded. "If it's trolled right."

The reef off Panther Point was showing its teeth. We ran parallel to it for several minutes until an eddy began to tug at the boat. Then a splatter of needlefish broke the calm—a sure sign of marauding salmon. Suddenly the pole dipped.

"We've got one!" Jeanne shouted.

The reel began to yowl. The pole jerked like a jackhammer as the fish lunged for freedom—two hundred feet, three hundred feet.

"What do you do now? I can't stop him!"

"Let him run," I answered, turning off the motor.

"You'd better take it, Dave. He feels like a whale—I'm afraid I'll lose him." And she handed me the jerking pole.

The battle lasted twenty minutes. We were exultant when we plucked a twelve-pound spring salmon from the net. He was a beauty with his silvery coat and meaty thickness. But it was beginner's luck; weeks passed before we landed another.

We barbecued our catch on the beach below the shack, over incandescent lumps of bark, basting him with bacon grease and parsley. Nothing in our lives ever tasted so delicious.

At dusk, we retreated inside. With the room curtained against the night, we sat under the Christmas tree and opened our gifts. Mine was a flannel shirt Jeanne had secretly made me. And there was also an all-metal hammer. "That's one you can't break," she teased.

While Kiki growled at a catnip mouse, I presented Jeanne with a box of her favorite lemon soap. When she opened it, her eyes grew moist. "From Robinson's, downtown Los Angeles—you darling!" And I felt her lips against my cheek.

In our letters home we had dropped a few hints, and now we tore open food parcels—canned hams, plum puddings, fruit. Even Bertha received a much needed flashlight.

Our celebration lasted until midnight. The floor was a sea of colored paper and ribbons, and the tree gleamed with birthday candles that cast dancing lights on the walls. We sipped cocoa and listened to radio carols, thinking how warm and cozy we were.

Jeanne squeezed my arm. "I can't remember a happier Christmas."

Toward midnight the radio battery died, so we wound up the phonograph and sang "White Christmas" accompanied by Bing Crosby. We couldn't pretend we were not a little homesick. I pulled out the last package from under the tree, an all-wool Indian sweater for Jeanne.

She slipped it on excitedly and stood before the mirror, stroking its fleecy softness.

"It's lovely, Dave," she said. "Now bring on the snow."

Winter obliged. As we fell asleep, snowflakes began drifting down on our island kingdom.

CHAPTER 12

Mutinous Cargo

WE AWAKENED TO A WHITE FAIRYLAND. When we looked out the frosty panes, firs were white obelisks in the tumbling flakes, and snow-laden drifts dotted the cove like miniature icebergs.

We had seen little snow before, only as it decked out the Sierra Madre mountains at home. After breakfast, we rushed outside. A hushed stillness hung over the cove. Kiki shook her paws, mystified by the white sticky stuff. The crisp air reddened our cheeks as we romped and threw snowballs like kids.

"Hope it lasts for days," Jeanne shouted, not realizing what she was saying. "Isn't it fun?" And a snowball whizzed by my head.

The next day ceased to be fun. During the night the thermometer dropped to ten degrees, and the next morning we dreaded to leave our warm bed. The door was stuck, banked with two feet of snow, and I was forced to climb out the window after wood. Icicles hung like frosty daggers from the eaves. The shed was barely visible. We tried to get a weather forecast. The battery lasted for the first part of the news: " War in Korea likely . . . maniac killer escapes from Victoria asylum . . . ten lost in apartment fire." Then the radio died for good.

By afternoon, we could not keep the trail open to the float. "Maybe we can reach Bertha at low tide," I said, hugging the stove. "She'll sink if we don't shovel her off." The weight of the snow was fantastic.

"And don't forget the roofs, too," Jeanne reminded me as she shoved another chunk of wood into Bessie. "We don't want Lii' Abner falling on our heads."

Fortunately the chicken coop was near by. We felt sorry for the three remaining hens, huddled together disconsolately. It was all we could do to keep warm ourselves.

We melted snow for water and drew the table closer to the stove. Outside, the cold lurked like an enemy, snapping the brittle madronas and threatening the firs, which leaned and sagged like tired old men. Birds haunted the doorstep for breadcrusts. Then one morning there were none. Like the woodpile, our supplies were dwindling. Coffee tasted bitter without canned milk and sugar, and there was only enough tobacco to make a few cigarettes.

After several days, the snow lost its novelty, and the room began to feel like a cell. We were cramped, always in each other's way.

"Good Lord," I complained, "why keep it so damn' hot in here? You want to suffocate us to death?" The heat was making me so groggy I couldn't keep awake to study.

Jeanne dropped her sewing. "It's not that hot. If you'd take off your sweater, you'd know it. We can't go around bundled up like Eskimos."

The cold magnified our imprisonment. I began to worry about my decision to tackle the guest cottage first. Was it fair to Jeanne? Her depressions were becoming more frequent. She had no privacy, that essential feminine requirement for putting herself together. It was impossible for us to bathe now; even the washline made the room seem colder. What was worse, with the bathroom sealed off we had to make periodic dashes to the little structure attached to the shed. Yet, I reasoned, wasn't our first step to make a living? I threw myself into my studies with greater vigor. On these, I was certain, the success of our dream depended.

When the sky cleared, the wind began. Polar gales whipped the channel feather-white, whirling the snow into five-foot drifts and tearing great chunks of roofing off the shed. The thermometer nose-dived until the fragile madronas cracked like gunshots. The cold penetrated the room deeper. Spilt water froze on the floor. We hung blankets across the windows and stuffed newspapers around the door, but even in bed, fully clothed, we were not really warm.

The mornings were the hardest. In the shivering dark, we stumbled over Kiki to light the lantern and build a fire. The kindling was brittle with frost. Bessie revolted. We had the choice of choking smoke or freezing cold. With the water bucket frozen, coffee took forever to brew.

The fourth morning began with a snarl.

"No oatmeal?" I glared at the bowl of cornflakes.

"There isn't any."

I pushed the bowl away.

"Don't be so fussy. Besides, I'm sick of oatmeal. We've had it every day for weeks." She nodded at the plate of soda crackers. "Those and coffee will have to do."

"Why didn't you put it on the last town list?" I demanded. "We're out of matches and soap too. What's the matter? You don't seem interested in anything any more."

Her face began to crumple. "I can't help it, Dave—it seems so hopeless."

"Don't worry.. we'll manage." I took her hand. "You've got to have patience—and faith, darling. Nothing is impossible until we've tried—given full measure of ourselves. The good Lord won't let us down."

"But he won't correct a foolish mistake," she sobbed. "We've misjudged everything—the money, the time, the foul weather, delays, the sea—everything."

She was admitting defeat, and that I could not bear.

"Listen, darling"—I lifted her face—"listen to me! If we're to make a go of this, you've got to believe in me. No matter what, we're going to be open for business June first even if Hell freezes over."

It took all our energy to keep comfortable. The lantern always needed filling, and Bessie had acquired a monstrous appetite for fuel. With our woodpile depleted, we scouted for bark at low tide. The urgency took away the fun of beachcombing.

To be rid of the snow, we would welcome rain again. "If it would only rain" became a chant that punctuated all our conversation. We were young and used to being active; there was work crying to be done. We were tired of sharing cigarette butts, and the sight of stale soda crackers made us cringe. At last, we were awakened by the sound of rain in the night, tapping the windows. We dozed off, thankful that relief had come.

In the morning, bedlam broke loose. When the frozen pipes began to thaw, fountains spurted in the bathroom and the plumbing hissed under the sink. Kiki leaped onto the bed as we rushed for towels and pots. The pipes leading to the sediment tank gurgled and hissed in the ceiling. Ping, pong, ping! went the pots. Newspapers began sliding off the walls and falling into soggy heaps.

"Shut off the water!" Jeanne cried, frantically clutching the basin pipe.

When I returned, she stood in the middle of the room glaring at the chaos. "If this isn't the damndest country!"

The North is a land of "too muchness." It's either too wet, too cold, or too dry. Fish, fowl, bugs, trees—*everything* thrives in this country—except people. You either hate it or love it. And to love it requires stoicism and strength. Relax your vigil, the greenery will swamp you. Life builds upon life in a brutal frenzy, climbing, killing, spreading, throttling for sun and space. Vegetation blocks your view, your steps, your shovel, and your drains. Trees shed their leaves only from sheer exhaustion. Summer is any warm day between two wet spells, followed by the remark, "We'll pay for this later." Moderation is unheard of. You're either shivering or sweating; exhilarated or depressed; thirsty or drowned. In summer, it's a struggle to have water in the house; in winter, it's a battle to keep it out. It takes a continuous effort to be comfortable, a herculean effort to be civil. To make anything your own is a real labor of love.

By late afternoon, the ditches were rivers overflowing their banks and the ground boggy underfoot. The island was like a ship rising from under a heavy sea. Water tumbled and roared into the cove, patterning the surface with veins of mud and debris. As the sun worked through the naked maples, the air became alive with the scent of moist evergreens. It was a joy to be outdoors again, listening to the music of kingfishers and feeling the sun's warmth on my shoulders.

After the shed roof was patched, I started on the plumbing.

"Think we'll have water tonight?" Jeanne asked.

"You'll have your bath tonight if you go after the mail and some grub. It's already nearly four—I can't fix these pipes and do that too." Nodding at the sea, I went on, "It can't get any calmer. Why don't you try it?"

There was still snow on Salt Spring, and we knew Myrt and Vic were well stocked with food.

"You know that engine bugs me. I crank my fool head off before it starts. Besides, the wind—"

"I believe you're chicken."

"I am not. I did it before, didn't I?"

"All right, I'll start it for you. If you want, you can leave it running over there. It doesn't take long to go up to the farm. Maybe Myrt will have some homemade bread."

That convinced her. She slipped on her Indian sweater and started for the door.

"Got some money?" I called after her. "We don't want to take anything from the Bettisses we can't pay for."

She took a two-dollar bill from the coffee tin. "You're funny. I believe you'd starve before you'd take anything from them."

I was foolishly proud. I didn't want them to know how bad off we were. While Jeanne was gone I mulled over our finances. The coffee tin held $6.42. We owed the store, possibly, $50.00. And across the last check stub were the words, "Funds Withdrawn."

I was tightening the basin trap when she returned, her arms loaded, her face grim.

"What's wrong?" I jumped up. "Did you hole the boat or something?" She was never one who could hide her feelings.

"Dave—it's the wharf. It's been condemned."

"Condemned?" I could hardly believe it.

"Yes. There's a notice posted on the ramp from the Department of Public Works. What can we do, Dave? We've got to have a base on Salt Spring Island."

I groped for an answer. "Maybe the local M.P. can do something?"

"What do the military police have to do with it?"

"Silly, that's what they call the local congressman. We'll write him—a General Peakes, I believe. Don't worry, honey. Canada needs American tourist dollars. He may go to bat for us." I put my arm around her, but she refused to be convinced.

"Dave, be realistic. There's nobody but us using it any more. Even Vic said so. They aren't going to rebuild it for a couple of foreigners—it would cost thousands of dollars."

It was then I noticed her hands. They were bruised and caked with blood.

"What happened? What did you do?"

"Oh, nothing." She shrugged. But in her diary that night she wrote:

> This has been a dreadful day. Dave talked me into taking the boat. I knew I shouldn't. At Fernwood, the stern line got caught in the prop. It killed the motor, and the chop drove the boat under the pier. To save her from the rocks, I tied the bowline to a piling. Cut my hands on the barnacles. Finally, I rotated the shaft enough to loosen the rope. With the first flip, the motor went—but I went down and the boat bounced off the pilings right and left. Fernwood is no place for a woman alone!
>
> Still more grief—Fernwood wharf has been abandoned. Can't see how Dave takes it so calmly. It's like a bad omen. We can't hang on here without it.

We had killed our last hen—poor Patricia—on New Year's Eve and tried to make it a celebration. We understood why she had withstood the cold. Her meat was tougher than cowhide—even Kiki had trouble gulping it down. It was a frosty moonlight night. The northern lights crackled, and spun amber and purple beams across the twinkling sky. We stood on the float and gazed in awe at the magnificent show. As midnight approached, we listened. The cove was pulsating with the splash of needlefish and the murmurings of ducks. Then we heard the far-off whistle of a distant lumbermill. How good civilization sounded to our ears!

The first mail after New Year's brought us a surprise—a G.I. Insurance dividend for $82.50. The check, a pretty green thing, reminded us of U.S. bills—it made us homesick to look at it.

"That'll buy lots of furnishings for the cottage," I said to Jeanne. "Or a new boat engine."

She shook her head. "Maybe sheep would be better. If the worst comes, at least we can eat them."

We ordered five ewes and a ram from a farmer near Fernwood.

Several days later at breakfast, Jeanne said, "Aren't we supposed to pick up the sheep today?"

"Good Lord!" I jumped up from the table, glancing at my watch. "I nearly forgot. The truck will be at the wharf in ten minutes."

"Need any help?"

"There won't be any trouble. I'll just tie their feet and pile them in."

That was how I figured it. But when you mix a city yokel with six sheep, a small boat, and an expanse of sea—you've got the makings of a heap of trouble.

On the way over, I began to feel uneasy. Do sheep bite, I wondered? The livestock manual didn't say. Of course, I had only skimmed it. They could, though! Suppose they kick too, probably like mules. Maybe I would need help. Come now, are you afraid of six little sheep? A small thin voice said, "I think I am."

A green truck was waiting at the wharf with its motor running.

"Sorry, can't help you," the farmer explained. "Got pigs expectin'."

He took my money, opened the tailgate, and the next moment I was alone, waist-deep in woolly gray creatures. Their bulbous eyes seemed to glare at me. For an instant I had the feeling I was surrounded by hostile Indians. "We're going to be good friends, aren't we?" I coaxed, vastly relieved to see that they had no horns. "Which one of you is papa? In my mind, the word "ram" meant unfriendliness. I looked under them, behind them—everywhere. Either there was too much wool or one had retractable sex organs.

To a symphony of "baa's," I herded my bevy of mutton to the end of the wharf and tied their legs. They bitterly protested the indignity, squirming and kicking. The smell was frightful, and I hurried—perhaps too much. Fortunately, the motor drowned out their bleating, but I was forced to sit on the transom rail to steer. So began a trip that is etched indelibly in my memory.

A hundred yards from Wallace, a tail got caught in the propeller shaft. The motor stopped as a horrible bellow of pain pierced the air. Then the revolt was on. In blind panic, the beasts broke from their ropes. Snorting, kicking, trampling each other, they began bounding over the side. One wide-eyed creature charged at me. I went over backward, arms flailing, with the grace of a circus clown. The next thing I knew, I was pawing the water while the sheep raced toward shore.

Jeanne met me at the float.

"What happened? Where are the sheep? And why are you soaked?"

I staggered from the boat. "The crazy critters mutinied." I motioned feebly toward the point. "They're over there."

Jeanne roared.

"It's not funny." I was annoyed. "They broke loose."

She went on laughing hysterically. "Didn't you know—the book says they're double-jointed?"

CHAPTER 13

Jeanne Gets Her Gun

A SPECK OFF THE POINT MOVED INTO THE COVE like a water beetle. It was a man rowing. We left our lunch and rushed to the wharf. A young man was standing on the float, looking tired and haggard, as though he had rowed a long time. He was unshaven, and his city clothes didn't seem to fit him.

"Have you any gas?" His tone was urgent.

The small inboard carried no gear or food, nothing except a rusty butcher knife on the transom seat.

While I was filling his tank, Jeanne said, "We're just having lunch. Won't you join us?"

The man glanced nervously at his watch, then at the boat. "No. I must be going."

"Come on. You look starved," Jeanne insisted, smiling. "It's all on the table."

"Oh, all right," he gave in, with a scowl.

We were puzzled by our visitor. He spoke with a slight lisp, and his dark, serious, deep-set eyes had a peculiar mystery about them. We tried to draw him out, but he steered shy of questions. That he was going north was all that we could learn. Conversation was confined to the weather. Kiki remained aloof, crouched under the stove, and glared at him.

We were relieved to see him go. When his boat rounded the point, Jeanne complained in disappointment that our first winter visitor had been someone who couldn't talk.

The next day, with a long shopping list, I set off for the village. We didn't want to be caught short again. The wharf was in miserable condition. Planks and sections of the railing had fallen into the sea. I held little hope for the letter in my pocket. When Roberta refused to start, I went up to Vic's for help. Snow was still heavy among the trees, but the road was clear. Vic was at the gate.

"Thought the winter had done yuh in, Tiny." He pushed the faded blue chauffeur's cap off his brow as sparks of devilment danced in his eyes. "Everybody's thinkin' yuh done high-tailed it back to th' States. Funny how rumors go, ain't it?... Say, yuh heard th' news?"

I shook my head.

"Yuh know that fella that escaped from th' asylum—that killer? Th' police boat caught 'im off the end of yer island. Lucky he didn't come in. Cut up the constable somethin' fierce, swingin' a butcher knife."

After that, Jeanne was suspicious of every boat. There was hardly a day that tugs, trollers, or seiners didn't rumble past. When one came near shore, she scoured it with binoculars and shouted from the window, "Dave, there's a boat going by."

I soon got so I ignored her calls. The cottage was going up slowly enough without interruptions. Mistakes constantly badgered me. Nipples I cut were never the right size, the wrenches I reached for never the right ones. If the lantern was not sputtering, it blinded me. Then there was the weather.

For a week now the thermometer had hovered in the twenties. The ground crackled underfoot and my hands, as I threaded pipe, felt like claws. The open studding did little to break the chilling wind. I longed to get the plumbing finished so I could put the siding on. While I worked, in my mind's eye I visualized the island in three years' time. The shack was a rambling bungalow with tile floors, flower boxes, and picture windows. The old mess hall, a lounge and store. Red-roofed cottages skirted the cove. There was refrigeration—electricity from our own generating plant—a washing machine for Jeanne, a bench saw for me. White shiny cruisers and sailboats were moored to the floats. People were swimming—laughing—playing—recuperating from the strain of civilization. The vision became my secret life and drove me on.

"Did you see that fishboat go by this morning?" Jeanne asked in a worried tone as we were eating dinner.

"No—what about it?"

"That's the second time it's gone by close to shore. Today, I could have sworn someone was studying the cove."

"Just beachcombers looking for logs," I surmised, unaware how wrong I was. "They won't bother us."

The lantern pushed valiantly at the darkness. The desk was a welter of books and plans. I was designing a gravity water system for the cottage. How many elbows? Tees? Where to put the filter barrel so it couldn't be seen? Did I have enough pipe? Would there be enough pressure?

"Please, Dave—come to bed," Jeanne called. "You can't go on like this night after night. You've got to slow down.... You didn't even finish your breakfast this morning."

"For heaven's sake!" I threw the book down. "Quit hounding me. I've got to push myself, or I'll never finish the cottage." I picked up the book and began to study again.

"At least you can be more sociable," she went on in a hurt tone. "We never do anything together any more—and you lose your temper so easily. Take the lamp, the other night. You never spoke to me like that before. You'd think I was your servant, not your wife!"

I looked up. "What did you say?"

"Oh, nothing...nothing." Her head fell back on the pillow. "You act as if you're in a fog lately. You don't like to talk, eat, or sleep. If I didn't know you better, I'd say you're in love with someone else."

That night I woke up suddenly, my hair bristling. Strange scraping noises were coming from the cove. While Jeanne slept, I slipped into my clothes and hurried out to investigate. Three Indians were knocking oysters off the rocks by lantern light and stuffing them into sacks. I crouched behind a log and watched them, wishing I had brought a gun and wondering what to do. Did I have the right to protect something that was not really ours? As I debated with myself something brushed my leg, and there was a soft meow. I lurched against the log. It rolled, thumping over the rocks into the water.

Startled, the Indians dropped their crowbars and sacks and looked up.

With expressionless faces they stared at me, their swarthy skins glistening in the flickering light.

Mustering my courage, I stepped forward. "This is private property. Go on now."

For a moment there was an ominous silence. Then they broke into insolent laughter.

"Did'ja hear that, Charlie?" the darker one said as he stealthily picked up his bar. "The man told us to clear out. What'cha think we should do about it?"

They fanned out slowly and crept toward me, keeping up their exchange of sneering comments. I inched backward, heart pounding as their malevolent faces came closer.

Suddenly there was an explosion behind me and the clam shell sprayed up at their feet. The Indians froze to the beach like grotesque statues. Then Jeanne emerged into the light, her automatic still smoking in her hand. Her face was hard and she held the gun with authority.

"You heard him," she shouted. "Get going before this thing goes off again."

Their faces looked as if a ghost had appeared. They turned, stumbling into each other, and ran for their boat.

Jeanne rushed over to me. "Thank God you're all right!... Do you think they'll come back?"

"I think they're more scared than we are," I told her, mopping the cold sweat from my brow.

The ordeal left us badly shaken. Not until we got back to the shack and heard their boat slowly sputter off in the darkness did we begin to relax.

"You sure came in the nick of time," I said over the hot coffee. "How did you happen to wake up?"

Jeanne looked thoughtful. "That's strange. I wouldn't have, if Kiki hadn't jumped on the bed."

Hearing her name, Kiki trotted over and began purring against my leg. I picked her up and hugged her. "You're one in a million Kiki-pooh."

A few days later, Jeanne confided to her diary:

> Another one of "those" days. Nervous headache and no appetite. Tried to make bread, but the flour was alive with wiggly maggots. Ugh! I'm really depressed tonight. This cursed cold and gloom make everything seem so hopeless. Maybe I'm just lonely. Dave's always so busy, busy, busy—he only knows I exist at mealtimes. We had our first serious quarrel yesterday. I knew it was coming. Every once in a while I have to let off steam. Afterward, I feel like a heel. Dave has so much to worry about. I wish I could get him to do something together—play cards, take a walk, listen to a record—or make love. We'd feel much closer. The way things are now, I don't feel needed any more—and I keep thinking something awful will happen.

Before January was over, the blow fell. It was as if the world had begun to slam the door on us. The mail brought a note that I read with a heavy heart.

> Mr. Conover:
>
> We beg to advise that your charge account has been terminated until such time as the balance of $126.66 is paid in full.

I shoved it into my pocket, hoping to keep it from Jeanne until her spirits picked up. But it was no use.

"What's the matter?" she said at dinner. "You aren't usually this glum."

I shook my head.

"It's bad news, isn't it?"

"I'm afraid so." And I handed her the letter.

Her face slowly whitened. "I knew it would come to this," she cried, crushing the letter. "And what do we do now? Sit here and starve to death?" She sprang up and went to the window.

"Don't get upset, darling. Things will work out. We'll manage." But my words only antagonized her.

"'We'll manage! We'll manage!' Is that all you know how to say? You crazy fool, it won't keep us from starving." Her voice was bitter, and tears streamed down her face.

I tried to hold her in my arms, but she jerked away. "We won't starve. There's plenty of fish and oysters and clams. Deer and mutton too." I fought to be convincing. "The days are lengthening now—we'll have a garden soon. Why, June'll be here before we know it."

"You don't understand, Dave," she said, her eyes pleading. "It takes money to live, to build, to pay for the island. Don't you see?"

"Darling, we've got to hang on. Things *will* work out. I *know* they will—if we just keep on fighting."

"Fine words. But they don't pay grocery bills, Dave." Her voice rose. "I'm interested in facts. Plain, solid facts. I can't live any other way, nor should you. There's no point in kidding ourselves. We've made a big mistake."

I didn't answer.

That evening, as we ate the cold stew, the quiet grew unbearable. "Well, say something!" Jeanne glared at me. "Isn't it the truth?"

"It's going to freeze tonight," I said, getting up from the table. "I'd better bring in the tools."

CHAPTER 14

Friday, the Thirteenth

AN UGLY SILENCE SETTLED BETWEEN US. It was as though we had embarked upon a project that shamed us. We could no longer face each other squarely, and our talk was limited to trivia. Each day Jeanne grew more edgy, her face paler as lines of strain deepened beneath her eyes, while I found myself more withdrawn and sullen, my jaw more stubbornly set. Meals were an unwelcome meeting. After dishes were done, we sank into our chairs like strangers in a boardinghouse, and the evenings were long with a bitter quiet.

"We can't go on this way, Dave," Jeanne said finally, coming to stand beside me. "Isn't it about time we had things out?" She paused, and I could feel her gathering courage. "Dave, we've got to sell the island."

It was as if she had struck me.

"Sell Wallace! Are you crazy? After all we've gone through—give up?" I shook my head. "Never!" I was stunned, for I could not bear an admission of defeat.

"Darling, be sensible," she begged. "We can't go on. You know that. We'll lose everything."

"Get this straight. We're not scuttling this ship. Now or ever—do you hear?" My fist struck the desk. "So put that crazy notion out of your head."

With a look of despair in her eyes, she turned away. When she spoke again, it was about something else.

But her words were like weights on my heart. We were no longer taking the same path. To dispel them I flung myself into work with greater fervor than ever.

January blew into February. The days were a blur of heavy slogging as, bundled like a polar bear, I put in the water systems and began installing the drains. Every day was a battle against wind, wet, and my own stupidity. Nothing ever went right. Pipes I measured and cut were either too long or too short. The hacksaw slipped and slashed my fingers. My back muscles throbbed from so much stooping. Days like these, I alternated between despair and frustration. To become a good plumber was beyond my wildest dreams.

Yet my days were full; Jeanne's were not. Housekeeping chores took barely half the morning—and, too, the weather discouraged her from going outside. "It seems so futile," she complained. "You haven't the time to go beachcombing any more, and I can't help you with the plumbing. What can I do?"

I thought for a moment. "You could make a fish trap. There's a spot on the north shore that's a natural—two parallel ribs of rock. All they need is bottling up."

What I failed to realize was the necessity of our doing something together. I never shared with her my problems of where to put the plumbing fixtures or my struggles to remove a stump. I blamed her depressions on the foul weather, and was too busy to imagine the terrible boredom that filled her days. At night she could not sew or read long before I'd feel her arms around my neck, hear her voice pleading, "Let's go for a walk. The moonlight is lovely, Dave."

Annoyed, I would fling down the pencil. "Not now. Can't you see I'm studying? How do you expect me to learn anything about fireplace construction?"

"Work or study! That's all you think about. I believe the island means more to you than I do."

"Don't be silly. You know that's not true."

Whenever we had one of these flare-ups, I'd assure myself she was just having another temporary fit of the blues. And usually I was too tired to prolong the argument, anyway.

A couple of days later when I went over after the mail, Jeanne came with me and we looked in on Vic and Myrt. I'm sure they sensed how hard up we were. They loaded Jeanne down with milk and winter vegetables, and she took everything they offered. I was so ashamed I almost boiled over on the way home. Someday, we would certainly repay them.

Often it was too rough to go after the mail, and one gloomy, drizzly day frequently followed another. After sixteen days with no sun, Jeanne's spirits reached their lowest ebb.

That night, breaking the heavy silence, she said, "You're so wrapped up in this island—all you need is a housekeeper. You never talk, never want to do anything together, never take me in your arms and kiss me. Even when you finally come to bed, you just roll over and go to sleep. Have you lost all interest in sex?"

"Of course not. I'm sorry—but I'm just too tired, that's all."

"Is that all you have to say—you're sorry. Why is it I have to blow up to get attention? Can't you understand how I feel? It's like being married to a stranger!"

I didn't know what to say to her.

"Well, say something—*anything*. What's wrong that you can't talk to me any more?"

"If you must know, every time I open my mouth you get all fired up," I blurted out. "No matter what I say, it makes you feel worse. Let's skip it." And I turned back to the desk in exasperation.

"That's not true," she argued. "You can't stand anything unpleasant—anything that upsets you. Did it ever occur to you that talking out problems keeps them from eating at your heart? You don't think of me. No. Only about yourself. I don't believe you love me any more." And she burst into tears.

I pulled her into my lap and pressed her head against my chest. Where should I start?

"I do love you, darling," I began, "more than anything. And I'm sorry I'm so crabby and self-centered and selfish. But when a person wants something so badly, everything else seems distant and small. Everything you say is wrong... what you do is wrong. But you don't mean it. It's a sort of madness, I guess.... I can't explain it."

I kissed her gently, and she sat up and wiped her eyes.

"I know what the island means to you, Dave, but it hurts to see you pour more of your heart into it every day. No matter how hard you work, though, it's slipping from your grasp—and you're killing yourself. I don't want the island to do this to you—to us."

"Didn't you say once"—I made myself smile—"that where there's a will, there's a way?"

She nodded.

"Then let's take each day as it comes, and not worry about tomorrow."

"I'll try, Dave."

She tried to invent indoor projects to fill her time. From old rags, she started to make a rug, but it frayed her nerves and she soon abandoned it. In her diary, she confided:

> Can't seem to keep my mind on anything. Must stop worrying so much. My latest project—to cover the rocker. Doubt if I'll have enough material or patience. Still dreary and raw. Even the sparrows have left us.

On Friday, things went from bad to worse. A sixty-mile southeaster had blown all day. There had been gales before, but nothing like this wild, hooting, shrieking wind that battered the island. Before dark, dodging falling branches, I checked Bertha again, feeling a bit uneasy and wondering if the wind would slacken, as usual, before changing to the west. In a westerly blow, I was afraid, the float might not hold.

The wind raged when we went to bed. "Think she'll switch tonight?" Jeanne asked.

I was almost asleep. "Doubt it," I mumbled. For she always died into a slop before she switched. So I thought.

Sometime in the night I awakened and sat up, startled, feeling the bed shake, the dishes rattle, and the windows drumming in the westerly wind.

"The boat!" I shook Jeanne, and we shot out of bed.

Leaning into the gale, the wind tearing at our faces and clothes, we raced down the path. At the gangway, I held up the swaying lantern. Waves were booming through the gap like charging white horses, and pounding the float against the bluff below us. Straining at her ropes, poor Bertha rolled and pitched broadside to the seas.

"It won't last long," I shouted. "Hold the lantern!"

I let myself down the dangling gangway by the cleats, and in two jumps was in the boat. Flinging out the anchor, I worked furiously to loosen the lines. The knots were tight, my fingers slippery . . . the last one wouldn't come. The pounding grew louder, and the spray blurred my eyes. I fumbled for my knife, opened the blade with my teeth, and slashed the rope. There was a sound of splintering wood.

"She's going!" Jeanne yelled. "Hurry!"

Frantically, I flew toward the gangway as the float split in half. It was then I saw Kiki—her luminous green eyes, her paws out four ways clinging to the shattered decking. I hesitated.

"Don't you dare, Dave Conover!" I heard Jeanne shout.

But it was too late. My feet hit the bobbing section Kiki was on. I felt it go under and the sea rushing about my knees as I swept her into my coat. Then I made a wild leap for the gangway. As my hands grasped the cleats, the float rolled out from under me. Waves surged round my waist, pulling at me as they subsided. The rock wall loomed above like a gigantic cliff.

I began to climb—one minute gripped by the sea, the next flung against the bluff. Slowly, painfully, I forced my body upward. I was afraid to look down. The wind struck my back like an icy compress, but at last I stumbled onto the bluff and collapsed in Jeanne's arms.

"You crazy nut!" she scolded while I blinked the salt from my eyes. "You could have been killed."

Then something gray and soggy squirmed from my jacket and rubbed against my leg, and I grinned. "We three have to stick together, don't we?"

The next day Jeanne was especially gloomy. The experience had frightened her badly. It was late afternoon before we rounded up the pieces of the float. We had just come inside for something to eat when she turned to me in tears.

"Oh, Dave—I wish we were home. It's too much for us."

I took her in my arms. "Sweetheart, I know how you feel. I'm still shaky and upset too—but as long as I can stand on two feet, we aren't licked."

There was a plea in her eyes. "It'll only be something else, Dave. Look at the well—your sickness—those close calls in the boat. Our luck has run out—don't you see that?" Her voice rose with a note of hysteria. "Dave, I'm scared... I've got the worst premonitions. We've got to sell and go home before something dreadful happens."

"Darling, listen—you must have faith. There's no other way."

I knew I hadn't got through to her; she was not listening. Her eyes were glassy with a far-off look as she wandered again in dim catacombs of fear. Then she broke from my arms.

"Don't my feelings count?" she demanded angrily.

"'Course they do," I stammered.

"Then how can you go on, knowing we'll end up hating each other?"

Words that made sense wouldn't come—I was too stunned. Never had I seen such a savage look in her eyes, such a contemptuous expression on her face. I wanted to hold her and kiss away the tears and tell her I loved her, but she threw herself on the bed, crying, "Go away, go away. Leave me alone!"

I retreated outside. All that I had strived for suddenly seemed to have no meaning. A terrible emptiness crept through me as I stood for an instant breathing the moist fragrance of the cedars. The low red sun glazing the apple trees and the precious stillness of the cove did nothing for my spirits.

I walked toward the wharf. *Was* the situation as hopeless as Jeanne thought? Should I go on—risk our marriage at the altar of a purely selfish, foolish ambition? The island was part of me. I couldn't give it up. I had to make Jeanne see—there was still time, still a chance.

I slumped down on a rock at the edge of the cove, closed my eyes and prayed: O God—am I wrong? Is it foolish to fight so hard for the island? Is it hopeless? I know I'm selfish—but I can't give up my dream. Lord, show me—somehow, some way—what we should do....

Before long, the cove was bathed in twilight. The firs were black lace against the crimson afterglow. The air held the sulfurous scent of low tide and the sound of whistling wings. Buffleheads and canvasbacks scooted onto the surface and began their guttural chatter in the shrinking dark. I began to shiver with cold.

"Chow's on!" Jeanne's voice rang loud and clear.

I stood rooted in my tracks. "Chow's on! Hurry!"

I ran up the path with Kiki bouncing ahead of me. The light was burning brightly in the window and the smell of frying food floated in the air. My heart quickened as I opened the door.

Jeanne had on her prettiest skirt and favorite green blouse. She rushed into my arms. "I'm sorry... I love you so much, Dave. I just had to let off steam."

I held her tight and kissed her lips and hair, feeling that life was good again and full of promise.

"You've got a right to be upset," I said. "Every woman wants security—and that's something we sure don't have."

"It's not that so much. You should have married a husky farm girl who doesn't mind getting her hands dirty or digging and scrubbing every day—somebody who wouldn't fret over—"

"You silly goose, *you're* the girl I love—"

Suddenly the room was filled with smoke.

"The French toast," Jeanne cried, swooping the frying pan from the stove to the sink. "That's all there is to eat," she said forlornly.

I winced inside, but tried to conceal my concern. "Don't worry, darling. I'll go out and dig some clams."

Our diet now depended mainly on seafood. Only parsley remained in the garden, and we both agreed mutton was the last resort. When we were out picking up bark we spotted flounder in the shallows, but it was too cold to use a line. Our breath hung in the air like wisps of fog. How could we catch them?

Jeanne considered. "I know," she said, and took off. She got her spear and we got our dinner.

A northwoods island in winter is not a good place to play Crusoe. It is difficult enough to keep the body alive, let alone the spirit. Cod had fled now to deeper waters and flounder were scarce. It was either too rough or too cold to fish for salmon. So it was oysters one meal, clams the next.

"Oh, what I'd give for a steak, fresh green peas, and a Waldorf salad," Jeanne said.

"Yes—even Patricia would taste like a feast," I agreed.

Then we ran out of tobacco. That really hurt. We experimented with dry maple leaves and barked at each other. Nothing satisfied the awful urge. When I went over after the mail, I found myself at Vic's hoping I could bum a cigarette.

"Quit the habit, son?" He arched his ponderous eyebrows.

"No, just forgot mine," I lied.

He smiled slyly and handed me his pouch. "Better take some with yuh. Life ain't worth a hill of beans without a smoke."

Food was a different matter. The Bettisses were kind, tactful, and generous. They pressed quantities of vegetables on us from their root cellar, far in excess of what we could afford. I grew more and more sensitive about this and about the fact that our visits always coincided with our need for something.

"We're not taking any more handouts," I told Jeanne when we got home one day. "Do you hear? Nothing we can't pay for."

She touched my arm. "What's the use, Dave? They know. After all, they're friends."

"They can talk, can't they? It's just what the Sultan would like to hear."

My thoughts were on the mortgage: the thousand-dollar payment due October first. No wonder he had demanded that the principal be paid back so soon. He knew this was a shoestring affair.

"You know something"—I stared at Jeanne—"I believe the Sultan cut us off purposely."

"Why?"

"To get the island back. He wasn't keen on selling it. Remember?"

"Oh, I can't believe that. He's just a shrewd businessman. Besides, the store has its rules."

The following day I patched the float together. It was a temporary job, for I was thinking of a larger, more solid, wharf. We had beachcombed two thirty-foot pilings. This was the only way of anchoring a float solidly. How we could get them driven didn't worry me. There were other things to do first.

To put the float in deeper water, we needed an approach slip—a sixteen-foot ramp out from the bluff. This would also hold the gangway more securely than roping it to trees. However, concrete footings on the sea floor were necessary to hold the posts. In winter, tides were high all day; the footings would have to be poured during zero tides at night. Timbers and planks were always plentiful for the gathering, and two sacks of cement sat in the shed.

All this I drew into a plan that night. Jeanne looked at it skeptically. "Are you out of your mind? Why, a pile-driver would cost a fortune!"

Truthfully, I hadn't given it much thought—I had such blind faith in the future. "Things will work out," I said, not looking up.

"'Things will work out'—what a dreamer!" Jeanne threw up her arms. "We haven't any money, and not enough food for a decent meal. Our days are numbered, and you know it!"

I ignored her and went on with the sketch. To argue would only make matters worse. The silence grew heavy. She loosened her hair and began brushing it nervously. The silence went on. Finally she flung the brush down and came over to me, her face taut, her green eyes glistening.

"Your cheeks are hollow, your pants hang on you like a sack, and you've lost at least twenty pounds. How do you think that makes me feel?"

"Good Lord," I exclaimed, "I've told you a million times that hard work never hurt anyone."

"When it makes a man look and act the way *you* do?" Her voice was shrill, and she went on and on as if she couldn't stop.

After a few moments I grabbed up a lantern and dashed through the door. The tide was nearly out and there were frames to build for the footings.

Each argument drove us farther apart. Other things, too, had widened the gap. I'd retreated deeper into silence, and I stayed longer at my work. I dreaded coming in. I could not counter her outbursts. Inwardly, I was far more upset and discouraged than I let her know.

That night her anger had subsided by the time I returned, and she got up from bed. "You look absolutely beat," she said. "Let me fix you something to eat. There's some left-over chowder in the cooler."

My hand touched her, and suddenly we were in each other's arms and I was kissing her.

We finished Vic's milk and went to bed. Jeanne slept but I couldn't. My mind was a bog of despair. I got up, lighted the wick lamp, and slumped at the desk. How had we drifted so far apart? True—I was selfish and often thoughtless... prisoner of a wild, unreasonable dream. I stared at Jeanne. Her hands were clenched, her mouth thin and tense. Her nails—so long and pretty and always kept with such exquisite care—were broken and ugly now. As the lamp flickered, her lovely face took on the pale pinched look of a corpse. I shuddered. How much more of this could she take?

For months she had been brave, taken every hardship. Now it was too much for her and she wanted to quit. Damn it! Why were women always so worried about the future, always wanting to play it safe? Living is not knowing, not being sure. Finally I opened the Bible. My eyes fell on a passage in St. Mark: "All things are possible to him that believeth." I read on, feeling suddenly replenished.

Next night, by lantern light in the bitter cold, I began pouring the concrete footings. It was devilish work. I mixed the soup on top of the bluff, then lowered it by bucketfuls into the frames on the sea floor below. It was a race against the tide... up and down the ladder... mixing... bucketing... lowering... tapping. And nothing went right.

Each night the sea washed the concrete away. Everything worked against me. The tides. The wind and waves. The lantern. I rigged up flares using oil-soaked rags. The rain put them out. And spoiled the mix. But I was determined to make the concrete stick.

Jeanne brought down mugs of chowder and we sat on the sawhorses under the flares. "You've been at this a week," she said. "Can't you see it's hopeless?"

"We've got enough for one more mix. It'll work this time."

At dawn, I was down on the gangway peering into the cold green water. The sea had been victorious again. Hardly a thimblefull of concrete remained. I would whip this problem if it was the last thing I did! I went back to the shack and checked the tidebook. There were only five or six more nights of zero tides till summer.

Jeanne put her hand on my arm, sensing what was going on in my mind. "Please, Dave, don't be so stubborn. The little float will do. Besides, Bertha is perfectly safe anchored out."

When I did not answer, she went gravely to the cupboard and took out the coffee tin. She counted the money carefully. We had $2.76.

"We need this for food, Dave. You can't draw cement out of the air."

That was what was bothering me.

Then it struck me. "We could catch a salmon," I said. "And take it in. Somebody will buy it."

We spent the rest of the day fishing. Luck was with us. By nightfall we had a seven-pound spring salmon. How it made our mouths water! It took all our resolve not to eat it.

"Why don't you go with me tomorrow," I said as I broke open the oysters for dinner. "The change will do you good."

The next day was Friday, the thirteenth. It was calm and crisp, not a cloud in the sky. As I wrapped the salmon in newspaper, Jeanne said, "Dave, I've got a feeling we'd better not go."

The disaster need never have happened if I had heeded her warning. Instead, I chided her for being superstitious, and argued that it might be days before we could get across again. Those footings had to go in while the tides were right.

In Ganges that morning the thermometer registered 29 degrees outside the store. The only signs of life were a tired blue pick-up truck full of logging gear and a horse-drawn wagon near the feed shed.

"Guess everybody is holed up for the winter," I commented as we went into the store.

"I'm glad," Jeanne said. "We look like anything but prosperous island owners."

Her green slacks were patched in two places, and my coat was out at the elbows. The island was rough on shoes too. Both of us wore soles from a beachcombed tire.

"Let's try the postmaster first," I suggested.

His eyes glistened through the wicket as we unfurled the newspaper. "Suppose you folks catch these all the time," he said. "Must be wonderful to have nothing else to do. How much do you want for it?"

We settled for sixty cents a pound. When the three sacks of cement were loaded in Roberta, we hesitated in front of the monkey cage. Our social prestige was at stake.

"I'll go in," Jeanne volunteered. "Coffee, flour, and sugar are worth more than pride."

Mine felt a little dented as I retreated to the soda fountain. A clerk rushed over. "Mr. Conova'?" "Yes."

"You're wanted on the phone."

As I followed him to the little office, I wondered who in the world it could be. "Hello," I said hesitantly.

"That you, Tiny?" The urgent voice was Vic's. "Better git here pronto, son. The sea is kickin' up an awful fuss and she's beginnin' to snow."

We rushed from town as the first few flakes began to fall. Moments later, we were engulfed in whiteness and crawling ahead blindly, numbed by the sudden cold. When we reached Fernwood, the sea was boiling and Bertha nowhere in sight.

CHAPTER 15

God at the Tiller

THE SEARCH ALONG THE SHORELINE WAS FRUITLESS. Weary and dejected, we piled into Roberta in the thickening snow and started back toward the wharf. It was nearly dark now. Overhead, wind-whipped firs swayed ominously like angry giants, while the wipers groaned from the load of snow. We drove on in a gloomy silence.

"Poor Bertha!" Jeanne said finally. "We'll never see her again. Oh, Dave, what are we going to do?"

"Don't worry, honey. We'll notify the police soon as we get to Vic's. Some ship will pick her up." I squeezed her hand. "We mustn't give up hope."

"I've run out of hope. This is the last straw, Dave."

When we reached our outdoor garage under the cedar tree, I put my arm around her. "We're going on," I said firmly, "as if nothing has happened. Someone will find Bertha. She'll come back—I know she will."

The sound of my own words comforted me, but Jeanne was not so easily consoled.

"What's the point of fooling ourselves? We can't go on without a boat—Bertha's gone for good. Let's be sensible and admit our failure before something else terrible happens."

For an instant I believed her. I looked out at the sea through the blur of swirling flakes. The waves were exploding over the wharf in a frenzy of foam and spray. Nothing could survive in that. Then I felt myself stiffen.

"We'll borrow Vic's boat and row over soon as the storm passes," I heard myself say. "Vic can drive down the road and signal us when he has news."

"Row over? Are you mad? We'd only be asking for trouble."

But my mind was made up. "We've got nothing to lose," I pointed out.

"Only our lives," she replied numbly.

We trudged through the snow to the Bettisses and stayed the night with them. They tried to bolster our spirits, but their cheerfulness could not hide the fact that there was little hope for our boat. They agreed to signal us when there was news. In the morning, we borrowed Vic's cockleshell and rowed to the island, silent and forlorn, pulling the sacks of cement on the little log raft behind us.

Every day we stared across at Salt Spring with mingled hope and dread. Jeanne grew increasingly despondent, and I buried my despair in slashing a trail to Pebble Beach to barrow back sand and gravel for the footings. The six-inch snowfall made the going tough. The ax dulled from striking hidden rocks, and balance was precarious in the rough slippery terrain. The second day I struck the beach. Then began the laborious hauling overland and pouring the concrete at night.

I hated to go inside now, for I could not face Jeanne. With mounting bitterness, she kept harping on the futility of going on. Stubbornly, I refused to quit—not knowing why any longer, feeling as if I had no control over the matter. Each night I prayed that God would help us—that Bertha would return. And asked for wisdom to make the right decision: Was I foolish to fight so hard for the island? Was it hopeless to try to carry on?

A week passed, then ten days. I felt my faith slipping. Why hadn't God heard me? Without Bertha, Jeanne was right—we could not carry on.

On the twelfth day, rain dissolved the snow and a cold gray mist floated down from the overcast sky. My shoulders sagged from the weight of my spirits, and my stomach growled with hunger. When I opened the door, Jeanne was sitting in the dusk, staring into space.

"Where's supper?" I demanded.

Silence. Then, "I'm not hungry."

I could not take another one of her spells. "You're not hungry! If you'd work a little instead of lying around moping and brooding all day, maybe you would be."

Jeanne sprang to her feet. "That's all there is to do on this blasted island—work! And for what? We're knocking our heads against a stone wall." Then she turned toward the window and went on in a lifeless voice, "Besides, there's no food."

Ashamed of my hasty words, I put my hands on her shoulders. "It's not that bad, darling. I'll dig some clams... and tomorrow I'll row over and get a few things from Myrt and Vic." The coffee tin still held a little silver.

Jeanne spun around angrily. "Can't you understand—God doesn't want us here."

I could not believe this. "He's just testing our faith, darling. Like Job, don't you see?" Standing there in the twilight, she seemed so lost and desolate. I wanted her in my arms.

"Bertha's lost!" she cried, recoiling. "Smashed to pieces. It's been twelve days... somebody would have found her. Don't you know when you're licked?"

"There's still a chance—we can't give up..." My voice trailed away.

"Give me one reason why not."

"We can't... that's all," I stammered. "It's the principle—"

"You mean pride. Damn' foolish pride!"

"There's only one way I leave this island"—my fist slammed the counter—"and that's flat on my back." After I spoke, I realized how foolish and selfish my words sounded. But it was too late.

Jeanne's eyes blazed. "Good Lord! I knew you were stubborn, but not a stupid idiot. This cursed island—you *do* love it more than you love me. All right, if you aren't pulling out—I am. Take your choice, Dave—which will it be: the island or me?"

I was stunned, then furious.

"You're running out on me? Haven't you got the guts to stick it out?"

There was a moment of ominous quiet as we stared angrily at each other. The silence was broken by the sound of a throbbing engine. We rushed to the windows. A long gray boat was creeping through the murk toward the gap, obviously coming in. Behind her trailed a launch.

When we reached the float we were greeted by a tall man in the green uniform of the fishery patrol. "I'm Officer Johnston," he said, and pointing to the stern, "Is that your boat?"

Nothing could describe our joy. As we gaped at Bertha, a feeling of reprieve flooded over us.

"Where did you find her?"

"Tree Island, just south of Nanaimo. Washed up on a clam spit."

"Unharmed?"

He nodded. "Not even a scratch on her bottom. How did she get away?"

"During that snowstorm," I said. "We thought she was a goner."

"Mister," he replied in amazement, "if she got away that day, the hand of God was at the tiller."

We thanked him and watched him leave. Then we stood, shoulders touching, and stared speechlessly at Bertha. The thought of her drifting through all those miles of islands and reef-strewn waterways without disaster or even any damage was unbelievable.

"It is a miracle, Dave," Jeanne said, misty-eyed. "You were right—God *does* want us here." A moment later, I felt her hand on my arm. "Forgive me—I've been such a burden, such a fool, darling. From now on we'll pull together. Nothing will stop us."

I kissed her and held her close while Kiki fussed at our feet. In the twilight, the three of us marched up the trail feeling as if we could live on clams the rest of our lives.

CHAPTER 16

Anniversary

THE ROOM WAS BRIGHT WITH MOONLIGHT. I could smell the scent of low tide and hear the sea chuckling against the rocks. For a long while I lay awake pondering. What a dismal husband I had been, thinking more about the island than about my own wife. Beside me, Jeanne slept peacefully, a half-smile on her lips and her hands unclenched.

I had erred miserably from the start in not building a home. My eyes swept the four bleak walls. The room was a prison. No wonder she worried about the future so much. The wharf would have to wait.

Next morning, I stretched the house plan on the desk: a ten-by-fourteen living room off the kitchen and a five-by-ten sleeping porch facing the cove.

"We can't do that," Jeanne cried.

"Why not? We've got everything—glass for those big screens of Lester's and roll roofing in the shed. And we can use firs for rafters, rocks for the fireplace, and drift lumber for the walls and floors."

"There isn't time. What about the cottage? We'd have nothing to rent—nothing to meet the payment."

"Hang the cottage! It's not decent living like this. You want a home, don't you?"

"Do I! I never wanted anything so much."

The living room walls went up quickly. Drift lumber was plentiful but in sizes that were difficult to use. We put the studding on the sides, spliced planks for joists, and ripped four-by-six's when two-by-four's were essential. By the time that I finished the roof, Jeanne had nearly sheathed two walls. She combed the beaches tirelessly, collecting her treasures by rowboat; then helped me saw and hammer hour after hour.

Pointing to the six-foot space between the windows, I warned, "That's for the fireplace. We'll tackle it when the weather is better."

Household chores suffered, and we took our meals on the run. The pot of low-tide goulash simmered constantly on the stove. The air seethed with its aroma; the memory of it sickens us to this day.

The March sun climbed higher and her smile grew warmer. We shed our sweaters. Gulls stood in pairs on the rocks, and the bush rang with the songs of towhees building their nests. There was a feeling of expectancy in the air as the island pulsed with the moist fragrance of spring growth.

From dawn until dark we labored, then laid floor by lamplight. The beach planks were rough and uneven. My plane scattered shavings ankle-deep, which Kiki explored like a jungle.

"Don't be so fussy," Jeanne protested. "Rugs will hide the cracks."

When the living room was completed and the garden spaded, we started work on the porch, using peeled firs and cedar posts. The frame went up in a day, the roof the following night. As we worked together, we felt the warm sense of comradeship flow between us again.

Jeanne's beachcombing adventures netted her a baby seal. Unfortunately, he lived only a few weeks.

A few days later when the porch screens were glazed and installed, Jeanne surveyed the finished rooms. "Looks like somebody is here to stay."

The couch, desk, and rocker sat in the living room; plates and pictures adorned the sea-aged walls. The porch was a compromise between bedroom and greenhouse. Our big maple bed fit snugly between the windows and the kitchen wall, so it was like sleeping outdoors.

We went to bed tired but happy. The air was full of sea murmurs as we watched the silver sickle rise, bright and fragile, above the firs.

Jeanne squeezed my arm. "Gosh, Dave, this is beautiful. Nothing could budge me from Wallace now."

During the building spree we had fallen to the rock bottom of living. The mudflats were pockmarked by our quest for food. The cooler held only a jar of mustard, vinegar, and a half-package of wheat germ. Each day something else gave out. First it was matches; instead we used lighters. Then it was razor blades. (My two-week stubble threw a damper on our lovemaking.) We rationed the lantern by using the wick lamp, and made only one trip to Fernwood a week.

When the kerosene was gone, we boiled tallow for oil and tried to ignore the smell. I didn't want Jeanne to ask Myrt for kerosene.

We brushed our teeth with seawater. Jeanne boiled it until it evaporated, for salt. We made soap from lye and tallow, following instructions in a camp craft book. Our clothes were a network of patches; every week a few more pants and socks turned into rags.

Poverty made us scavengers. We routed herons and kingfishers from their fishing grounds in our search for cod and flounder, and drove mink from the cove by stealing their crabs. These little gluttons tackled crabs larger than themselves, bringing them up from the deep to eat on shore. In our hunger, we ignored pangs of guilt as we devoured their prize catches.

Our diet grew dreadfully monotonous. We dreamt of sumptuous dinners . . . smorgasbords . . . that we were gorging ourselves at Perino's. We were so haunted by visions of steak, oranges, sugar, and bread that we began to wish that a ship would run up on the rocks.

"We're starving, Dave," Jeanne moaned. "There must be plants and roots we can eat." But after a bout of intestinal trouble from mushrooms, we were leery of anything wild but dandelion greens.

Then I had an idea. "We'll write the library. Maybe they'll send us something—even a book on mushrooms would help." Food so filled our thoughts that Jeanne mentioned it in her diary almost every day:

> What a godsend! Cabbages and marrow from Myrt today. In exchange, Dave gave them a pail of oysters. But we miss coffee most of all. We lived on it. Saw the sheep this afternoon. They're so innocent and friendly—one ate right out of Dave's hand. I don't know how we could kill one.

And another day:

Fish trap empty again, so tried for flounder all morning. No luck. If we could only find more to eat! I worry about Dave—he looks so lean and haggard. He's going on sheer nerves.

Finally, Jeanne said, "Dave, we can't go on like this. I can put up with that... that goulash, but you need meat—proper food to do heavy work."

"I know," I sighed. "I hate the stuff, too."

"Why don't you shoot a deer?"

"What's wrong with mutton?" I asked, surprised.

"Nothing, but wouldn't beefsteak taste better?"

"Beefsteak?"

"The sheep will lamb soon. If we hold off, they'll bring twenty-five bucks apiece, Vic said. That'll buy a lot of T-bones, won't it?"

I had to agree. Two months was a long time to go without a taste of meat. Even then, the hamburger had been watered down with oatmeal and cracker crumbs, A juicy venison roast would do very nicely while I waited for the T-bones.

But when I plunged into the forest with the rifle in my hand, I began to feel differently. I forgot the dull ache in my stomach as gooseflesh crawled up my arms at the thought of becoming a cold-blooded killer. Sweeping aside the salal, I told myself not to be silly: sentiment is fine on a full stomach, but not on an empty one. Killing for food is a practice as old as life.

I pushed stealthily through the grove of ancient firs. The ground was soggy, the air raw and heavy with the scent of decaying leaves. Crows flitted from limb to limb, cawing in protest. My throat was dry, and again I felt the gnawing ache of hunger. As I stepped into a clearing, there was a sudden rustle of leaves. Not a dozen feet away a deer bounced from the thicket and stood motionless, studying me with its big, pensive eyes. It was a fat two-point buck.

I shouldered the rifle slowly and took aim. The great limpid eyes kept gazing at me, innocent, curious, and trusting. My finger gripped the trigger... it wouldn't budge... the barrel began to wander. Lowering the rifle, I whispered, "Go on, boy. Beat it!"

With legs like springs he bounded into the bush. As I turned toward home, my eyes fell on the gun. I grinned. The safety was still on.

On Friday, the library came to our rescue. A book has been known to change a person's life, but *The Guide to the Edible Wild Plants of British Columbia* really saved ours. To the staff of the Open Shelf Library* of Victoria we give our humble thanks.

From that time on, Jeanne spent her days in the woods, then over the stove, in a frenzy of experimentation. Our bizarre first meal she recorded in her diary:

> For dinner tonight we had a crabmeat cocktail, wild onion soup, boiled fiddleheads, and a tossed salad of watercress, dandelion leaves, chopped rosehips, and mushrooms. I cooked the fiddleheads (bracken) in seawater for twenty minutes. They taste like asparagus to me—Dave says, artichokes. Bracken is as plentiful as salal. We'll try roasted camas bulbs, shepherd's purse, and pigweed tomorrow. Can't wait to locate chicory. Dave's sure he's seen it. Then we'll have coffee—hurrah!

Now, as I look back on those greenhorn days, I realize we never had time to be really afraid. Once into an ordeal, you're too busy making the most of it. Fear is far greater at the onset. We found, too, that everything has its compensations. Though hours, even days, were spent on tedious tasks, never a moment was wasted. The advantage of small jobs is that you can think about larger ones. In rooting out a stump, I solved the footing problem—open-ended oil drums allowed the concrete to set. And as I mixed concrete, I figured a way to make the porch windows slide back and forth. Nothing in life comes along that a man cannot bend to his purpose.

* A Provincial Government service that mails books postage-free to those living outside the reach of city libraries.

Then it was April. Sunshine. Crystal blue days. Overnight the maples budded, the grass greened, and the fiddleheads unfurled into giant bracken. The bush was a symphony of bird songs and full of bewildering smells. Lambs frolicked in the upper meadow. Each day was more beautiful than the last. How I missed my camera!

We had never felt so confident, so healthy, and so hard. Every hour was utilized—finishing the cottage, planting the garden, crusoeing for drift lumber, or raiding the forest for food and poles.

There came a morning when hammering on Salt Spring signaled good news. When we went over after the mail, a pile driver and crew of five were dismantling the wharf. A lumber barge stood expectantly offshore. We couldn't wait to question the huge foreman.

"She'll be finished by Easter." He nudged up his steel helmet. "Else I'll lose my crew. The boys are anxious for a holiday." His voice was as hard as his features, but his blue eyes were soft and gentle.

We told him he was welcome to tie up in our cove and that we'd keep an eye on things.

"Might just do that," he replied.

On Friday afternoon, a week later, the noise abated on Salt Spring and we suddenly found ourselves with a pile driver, roped from shore to shore. It seemed strangely out of place, its giant tower looming above the maples, like an oil derrick in the middle of the cove. It monopolized our conversation, the responsibility for such expensive equipment weighing on our minds.

The day after Easter brought good fortune. In return for caretaking the pile driver, the crew drove our piling. They had dropped the ropes from shore, and the tug was ready to depart with its tow. I was lining up the float logs, wondering how to join them without spikes or drifts to hold the stringers, when the foreman noticed my predicament.

"Looks like you need help," and he turned to his crew. "Okay, boys, let's give him a hand."

In no time, the two little rickety floats took shape. We thanked the men profusely. As the last glimpse of the pile driver cleared the fir tops, Jeanne summed up our feelings: "It's wonderful how kind and thoughtful such rough-looking men can be."

The days fled by with the sea like plate glass. In the rowboat, Jeanne's world stretched out. She discovered banks of trilliums and bluebells, caches of flagstone, and with increasing regularity she caught salmon and cod. Fish became our main meal. She tried every way of preparing it—baked, boiled, fried, casseroled. We ate so much cod our arms began to feel like fins. She noted in her diary:

> Cod is an ugly fish—huge spikes along his back, thick scaly skin, and big bulbous eyes. When he strikes, you think it's bottom—you have to winch him in! His mouth is so tough it takes pliers to recover your hook. He's 90 percent apparatus, 10 percent food. You just strip the flesh off his back. It tastes like sole, and is not as tiresome as a steady diet of salmon.

To augment our austere diet, we took fish and oysters to Salt Spring Island to sell or to trade with the Bettisses for vegetables, flour, milk, and eggs. I was ashamed of our poverty—our shabby clothes, my mangy beard; Jeanne had the courage. I was afraid people would think we'd gone bushy and that we were broke.

"I know how we can raise money," Jeanne suggested. "Your photographs. There's a folder of them in the trunk."

She got them out and began going through them excitedly. "Look at these landscapes. Death Valley. Capistrano. They're lovely, Dave. Some of them ought to sell."

Jeanne had an uncanny way of sensing which of my photographs would pay off. My strength was in the creative aspect of photography; for the selling, I nearly always depended on her.

From the dozen prints on the table, she selected six and sent them off to magazines.

The next morning dawned with such pure enchantment and brilliance we could hardly believe our eyes. Forest, sky, and sea were their deepest hues—still, peaceful, and shining.

"Let's play hooky and explore the islands," I said from the window. "I bet we can't find one as lovely as this."

With a picnic lunch, we took off for the Secretary Islands, our nearest neighbors to the north. On the chart, they looked like two gumdrops stuck together, roughly the size of Wallace.

Bertha hummed merrily. The sea air whipped our faces and made our blood tingle. As we neared the islands, a tiny channel opened between them and the water grew shallower. Crabs and cod scurried among the reeds. It seemed as though the greenery was wilder, even more luxuriant, than on Wallace. Then the channel widened into a small bay with a white shell beach. Behind it, engulfed by the forest, was an ancient house, an orchard, and some scattered outbuildings. We anchored and waded ashore.

Once on the beach, we were struck by the haunted feeling of the place. The air was heavy with decay and mystery. Through thick undergrowth, we groped toward the house. Scavengers had stripped it bare. We peered into the windowless openings, surveyed the yard from the tumbledown porch.

"Somebody sure did lots of work here," Jeanne remarked. "Judging by the landscaping, traces of fence, clearings, and barns, he must have been quite self-sufficient." We had the uncomfortable feeling our presence was an intrusion, and it made us sad that Nature had defeated this man's Eden.

Since time immemorial, man has suffered from island fever. It is a highly contagious malady, and once contracted, there is no hope of recovery. Everyone has felt its symptoms... the longing to live simply, quietly, fully, on a dab of land surrounded by sea, sky, and trees where even the air you breathe is all your own. But man—the supreme social animal—usually escapes his bondage only in dreams. For the comfort and security of his rut always outweighs the quest for freedom, which can be realized only through a marriage of resolve and sacrifice.

We gathered our customary loot—age-old boards, ornate bottles, spikes, and rusty pans. There was hardly a whisper of air, and the sun scorched our backs. With our treasures collected on the beach, we shed our clothes and raced into the water to cool off. Afterward, we sprawled on the clean, warm shell and ate our lunch of flapjack sandwiches and chicory coffee. Jeanne had ground the flour from dried camas bulbs. Mixed with milk and fried, it resembled a pancake, and smothered with rosehip marmalade, it was a gastronomic triumph.

From the overhanging madrona there was a sudden peal of castanets. Freddie dropped at our feet and strutted stiffly to and fro like a wound-up toy while we plied him with crumbs.

For a long time we sat drinking in the scene, inhaling the tangy aroma of seaweed drying in the sun and watching a gaunt heron stalk the shallows with patient dignity. Everywhere around us the jagged firs echoed with songbirds.

Jeanne broke the silence. "What enormous firs. I'm jealous. I believe they're larger than ours."

The trees that rimmed the basin—firs, cedars, and hemlocks—were like a great wall of green pushing back the sky.

"Maybe," I conceded. "But this cove doesn't match ours. Or Princess Harbor. And we have more fruit trees and cleared land." The thought that someday we'd own these islands never entered our heads.

The rest of the day we netted crabs, gathered chicory root and pigweed, and combed the shoreline for lumber. As the sun faded we trolled home, hoping for a salmon. But our record for the week remained unblemished.

In May, boats grew more numerous. People escaping the cities and the heat were taking to the sea. Yachts would gas up in Ganges, where they'd mention us as "points of interest." Often they idled by the cove, wondering if they dared come in. We were glad the reefs scared them off. The way we looked—like shipwrecked waifs—we weren't anxious to meet people in fancy boats. However, compared to Canadians, American yachtsmen weren't nearly so cautious.

Once when we were weeding the garden a yacht entered the cove. Smeared with dirt, our shorts soiled and frayed, we were a sorry-looking pair. As the huge flying-bridge cruiser swung toward the dock, Jeanne reached for her blouse and I nervously fingered my unruly beard.

A wonderful sight made our hearts throb. From the stern flew the Stars and Stripes.

"Americans, Dave!" Jeanne grabbed my arm. "To think they've got steaks . . . potatoes . . . real coffee and cigarettes on board."

"And Hershey bars," I breathed. "And razor blades, and people with the latest news of home."

We waved from the float in welcome.

The sleek fifty-foot cruiser came toward us, its decks aflutter with people in bright clothes. The man at the wheel in blues was nursing a glass. As he spoke, he surveyed us arrogantly. "Couple of queer-looking nuts. We'd better move on."

A woman in a tailored white beach outfit agreed, "Yeah, these islands are full of 'em." And the yacht shot from the cove.

Jeanne and I stared at each other in shocked surprise. The dirty so-and-so's, I thought. Even a savage wouldn't have any worse manners. I remembered Colin, his gracious ways—the hospitality of his cramped little sailboat. Here, now—our own countrymen!

Jeanne was more philosophical. "We're happier at least, and we don't have to drown our boredom in drink."

As May wore on, work on the cottage came to a standstill. Again, shortages plagued me: finishing nails, plywood, one-by-two's for the cabinets. I had misjudged badly. Each day became more frustrating as I ran out of more and more things: masonite flooring, hinges, cupboard latches. The hole for the cesspool was blocked by boulders. Time was pressing—and I had lost my only helper. Between housekeeping and scouting for food and drift, Jeanne's days were full to the brim.

"How's the cottage coming?" she asked when we finally sat down for lunch.

I didn't have the heart to tell her, but—as usual—she seemed to sense my troubles just the same.

"It's strange how our fears change," she said thoughtfully. "First we worried about how we'd get through the winter. Then, how we could manage without money, food, or a boat. Now, if we're going to be ready in time."

"It's only three weeks till June. I don't see how we can do it."

Jeanne seemed to be turning over something in her mind before she answered. "We can—if you can stand unmade beds, dirty clothes, and clams again."

From that day on, she was by my side—balancing Gyproc above her head while I nailed it to the rafters, filling nail holes with putty, and helping me hoist boulders from the hole. Her help heartened me, for the situation called for all the powers we possessed. I was slow and overly meticulous. Her quickness and realism prodded me on. And, too, I admired the resolute way she tackled the smallest chores and the effortless grace with which she packed lumber from the shed or brought bricks up from the beach.

Finally, all at once things began to go our way. A check for the photographs gave us our first good meal in months. Two sets of twin lambs greeted us in the meadow. We were eating spinach, radishes, and lettuce from the garden. The fish trap held cod, and the shallows crawled with flounder. And for helping Vic wrestle out the two stumps in his yard, he presented us with six banties, two guinea fowl, and a sack of netted gems.

"You know something," Jeanne said at dinner. "I think we're here to stay."

In late May the sale of the lambs restored our village credit. It was boat day. The *Princess Mary* had just docked and the store was packed. I shoved the butcher's receipt, a pretty pink thing, in my pocket and approached the hardware counter. The Deacon's cadaverous face looked gloomier than usual.

"Anything wrong?" I asked cheerfully. "You act like I'm bad news."

"You are, Mr. Conova'." His spaniel eyes gazed at me. "You are indeed. I lost ten dollars you folks wouldn't last the winter."

My curiosity was aroused. "Who was the smart fellow?"

"Vic. He was anxious to bet with everyone, and ten-to-one odds did seem pretty safe. He's done cleaned out the store."

The old scalawag, I thought. He had deliberately spread stories to increase the odds. No wonder the villagers had treated us so coldly.

The Deacon's eyes softened as he leaned toward me. "But you know something, Mr. Conova'. I'm kinda glad I lost."

While he was weighing ten pounds of nails, I felt a hand on my shoulder. There stood the Sultan with a dour expression, his huge shoulders humped like a majestic eagle's. His dark, deep-set eyes had a frightening intensity as he studied me over his spectacles, as if I were a wayward son.

"Yes, sir," he said. "A fine lot of lambs the butcher tells me. Sorta pulled out your chestnuts, aye?"

I was tongue-tied. "Yes... lovely chestnuts, I mean..." And I felt my face go red.

After an exchange of pleasantries, I dashed from the store, eager to get home to break the news to Jeanne. Outside, Vic was loading mash into his truck. When he saw me, he flew into the feed shed. I cornered him behind a bale of hay.

"You old rascal!" I seized him by the collar. "How much did you make?"

"Eighty dollars," he gulped. 'Don't get mad, Tiny. Jest tryin' to make enough for a new brooder."

I released my grip. "Fine friend you are." I couldn't keep from grinning. "Why didn't you let me know? I could have placed some bets too."

That evening while the T-bones cooked, Jeanne fondled the pink receipt. "Just think! No more clams, fiddleheads, or monkey cages. Now we can act like what we're supposed to be—rich, nutty Americans!"

We hugged each other and laughed.

The next day we sent off a classified ad to the Seattle papers. And held our breath. On this, we thought, hinged the fate of our dream:

> VACATION ON A PRIVATE EVERGREEN ISLAND. HOUSEKEEPING COTTAGE, FURNISHED, $75.00 WEEKLY. FISHING, SWIMMING, LOAFING. WALLACE ISLAND RESORT, GANGES, B.C.

After a few suspenseful days, we were swamped with letters. Some with reservations. All wanted more information. What was the size of the island? The water temperature? Did we have baby-sitters? Pasteurized milk? Good swimming beaches? An island vacation was appealing. Reservations poured in. By the end of the week, we were in a state of high elation. The cottage was booked from June 6 through Labor Day.

But our troubles were not over. The well was dropping fast—and we had only two weeks to finish the cabinets, the built-in beds, the shower stall, and to test the water system. Two weeks to put in the missing windows, install a stove, varnish the knotty pine, and paint the boat. Night and day we worked like demons, struggling with windows that didn't fit, fighting leaky pipes and a toilet that insisted upon flushing up in the shower drain. On the third day, our hopes dwindled.

Discouraged, we trooped into the house at noon and made some coffee. Neither of us was hungry. As we sipped the coffee, we were too preoccupied with our dilemma for speech.

"Damn it!" I burst out finally. "Even if we had a month, we couldn't do it."

Jeanne's green eyes met mine. "That's not like you, Dave. There must be something we can do."

I was too weary and depressed to think.

All of a sudden she brightened. "I know—why don't we move into the cottage and rent this place instead?"

I considered that. "Sure you want to?"

It would be rough—a roof over our heads, that was about all. There was no furniture or place to put clothing. No shower or hot water. The toilet didn't function. We'd have to sleep on cots. Yet we had to appear clean, neatly dressed, and pleasant to guests. Could it be done?

"I'm positive," Jeanne said. "Let's move in right away, darling. It'll give me time to get the house—I mean, our deluxe cottage—ready."

"What about heat—cooking?"

"The fireplace will do. And I'm used to getting meals on the camp stove." That the fireplace was untested and the chimney half-completed didn't matter. "It'll be hard, but we'll need every cent we can get, come September." Our mortgage payment with interest was $1,420—payable October 1.

On our first anniversary, we took inventory of ourselves. Outdoor life agreed with us. We were sun-bronzed, lean, and hard. And our hands were rough and strong. Giving up smoking had doubled our endurance and quickened our perception. We could hear a fir cone drop a hundred yards off and clearly see the eyes of a seal across the cove. What was more important, we felt vitally alive. For the first time, May was something we felt in our bones, not merely a fleeing abstraction on the calendar.

Island life bore fruit we never expected. We had become better acquainted with ourselves. I had discovered that good teamwork came from a willingness to compromise; that perfectionism only frustrates—for it's better to have two jobs reasonably well done than try to do one too perfectly. Above all, we reveled in the glorious feeling that comes with achievement. Each problem we overcame not only tightened our grip on the island but widened our ability to act, freeing us from the self-imposed shackles of fear and doubt.

After dinner, we sat on the edge of the cove in the twilight and mused over our old way of life, which seemed like a thing of the far-distant past.

"It's a wonder how we ever stood the city," Jeanne said. "We've got everything here, and we're contented as two people can be. I don't think we could ever go back."

"You're being romantic," I chided. "Don't you miss dancing, the movies, those parties we used to go to?"

"No, not really. I don't think we're missing anything worthwhile."

We never tired of looking at the cove. The tides shaped it differently each time, like an ever-changing hourglass. There is both excitement and replenishment in nature's ebb and flow. People are happier—more natural and hopeful—who see the sun rise and set, a hillside leaf and flower. Exposure to nature's rhythm and beauty quiets and fortifies the spirit. I could see this inner calm in Jeanne's face, and I could feel it in the sureness of my own steps. Just as the cove tapped a larger body, we felt a part of something far greater than ourselves—an order, a giant plan, in which we were essential participants. The beauty and harmony that surrounded us were too perfect to be a product of blind chance. It was the handiwork of a supreme architect.

We loved to row in the moonlight. After a while we pushed off in the cockleshell. It was our magic carpet, for at night the water seems but thicker air. The oars glowed with sparks of fire and the warm, resinous smell of cedar filled our lungs. We were silent, listening to the creak of the boat, the muted voices of ducks, and the far-off tinkle of cow bells. The moon hung like a silver bubble above the gaunt firs. Around us we could feel the serenity as if it were a tangible thing.

We stopped rowing, drugged by the wonderous night. Jeanne embraced the cove with her eyes. "It's hard to believe all this belongs to us—it makes me feel selfish. Most people have so little. It doesn't seem right we have so much."

I thought for a moment. "When one shares his dream with others, nothing is too much. God gives him a little bit more."

CHAPTER 17

"And Not a Drop to Drink"

THE CRISIS CAME SWIFTLY. The well was dropping fast, and the siphon to the house kept breaking.

"It doesn't look good," I said, bringing in a bucket of seawater to flush the biffy. "We shouldn't be running short this early."

"Why don't we dig the well deeper? Maybe more water will come in."

I shook my head. "Not a chance. She's down to bedrock now. All we can do is pray for rain."

Since March, very little rain had fallen. The forest exhaled the hot odor of fir needles. The ground felt like concrete, and the madrona bark crackled in the thirsty shade. Lettuce and spinach, once so luxuriant, were now wilted and dying. The sun's stinging rays forced us to wear hats, something we had never done in California. We found ourselves constantly squinting up at the cloudless sky, wishing, praying, for rain—even a thundershower—to save the garden. And, too, we worried about the sheep, knowing they had no water.

The heat made the house unbearable, and we fled outside to eat under the maples. Each day the garden took all the well produced, except for a pailful to drink and cook with. One morning, when I dropped the bucket into the well, it made a loud clang. The well was dry.

When I rushed to the house and told Jeanne, we stared at each other in dismay, like two frightened youngsters. Without water, we knew Wallace was not only useless, but impossible to sell. Two choices faced us: to dig or to drill. The first offered little prospect of success—certainly not of finding enough water to supply a resort. If we drilled, the chances were better, but this we could not afford.

The Gamons, our first guests, were due in nine days, so I had to act fast. Actually, my mind was made up—I really had no choice. This was a job for water drillers. They were in the business to find water. But, here again, I was wrong.

Fortunately, Vic had water to spare. That afternoon, I brought back a full drum and we doled it out to our thirsty plants like liquid gold. The water tasted oily, but it saved the garden for a while.

"What are we going to do?" Jeanne asked at dinner, breaking into our depression. Her face looked small and pale in the dusk.

"Drill," I replied.

"At ten dollars a foot? Are you serious? All we've got is eighty-five dollars in deposits. Why, it might run into thousands."

"It's the only sure way. We've got to—"

"You're crazy! Where will we get the money?"

I got up and stared out the window. "It's too late—I've already ordered the drilling rig. It'll be here in the morning." Over Vic's phone, I had frankly explained our plight to Columbia Water Wells, and they'd agreed to give us ninety days.

Jeanne's exasperated sigh was clearly audible. "I hope the good Lord sticks with us. We'll surely need him."

The landing barge arrived next day at noon. A bulldozer rolled out; then the drilling truck groaned onto the beach. We were appalled at its size—an old army Mack still painted olive-green, with a midget Eiffel Tower collapsed on its back. The driver-operator, a handsome lad with black hair and blue eyes, leaned out the cab.

"Where do you want the hole, Mr. Conover?"

I was astonished. "*Where?* Gosh, how would I know? Isn't finding water your business?"

"Nope, it's the diviner's," he said. "We just drill the holes."

"You mean one of those guys who walks around with a willow twig?"

He nodded.

This was the last thing we had expected.

Jeanne put her hand on my arm. "Why don't you go over to Salt

Spring? Maybe Vic knows of a diviner."

To my surprise, nearly every Salt Springer over sixty-five professed to have the occult power of water witching.

"Jest a matter of pickin' the nearest," Vic declared, leaning on his pitchfork. "Try Featherstone, down the road. He did right by us. Yuh can't miss the mailbox. It's the one on the stump."

The stately old islander looked like an El Greco figure—gaunt, long limbed, and balding. And he was eager to display his powers. An hour later, in strained silence, he walked with us up the path toward the dry well. The forked willow stick in his hands began to bend.

"Right here." He stopped, his gnarled fingers straining against the mysterious earthly pull. "There's a vein." His authoritative voice inspired confidence.

"How deep?" I asked eagerly.

Silently he counted the dips of the willow. "Twenty-eight feet."

I felt a flood of relief. We not only had water, but at a fairly shallow depth. Yet, I had an uncomfortable feeling this was all too simple. On an island, nothing comes easy.

"At least it won't be as expensive as we thought," Jeanne said.

When the bulldozer had cleared a road to the site, the barge departed. We shivered with excitement as Bob raised the tower and began to drill.

The following days were the most suspenseful in memory. To our horror, the drill required fresh water. We had to haul over several drums a day and, panting and sweaty from the heat, roll them up to the site.

Ten feet down, the drill struck sandstone. Thump... thump... thump! it went all day, like a huge hammer striking the island. Progress was slow, a foot an hour. Each thump meant twenty cents. We were up at the rig every possible minute, watching the cable tighten, listening to the clank of machinery, feeling the ground shake—and hardly daring to breathe. Or to work or eat.

The second day brought disappointment. At twenty-eight feet, the hole was dry. Thump . . . thump . . . thump, the pencil-shaped drill drove deeper. Fifty . . . sixty . . . seventy feet.

"What are the chances, Bob?" I asked at seventy-five feet. He shrugged. "When you're drilling, there's always a possibility."

Our visits to the rig declined. We were sick with apprehension. Then, one morning while we were moving into the cottage, the thumping suddenly changed to a dull thudding.

"Water!" Jeanne shouted, dropping her armful of pans.

We raced up to the rig. Bob had just brought up the bit, and it was black as pitch.

"Coal," he said.

Even if it had been oil, we wouldn't have felt any more enthusiastic. Then I remembered reading how coal mines were often flooded with water, and I said, "Keep drilling, Bob. It might be a good sign." But I had guessed wrong again.

The vein was shallow. By afternoon the thumping returned, and I knew we were once more in sandstone. When we approached the hundred-foot mark, the suspense grew unbearable.

"We've got to do something," Jeanne said. "The hole is a dud. We just can't sit here. Let's get another diviner."

"All right," I agreed. "Maybe Lester can help. There ought to be a good one in Victoria—if there is such a thing."

Lester came through like a prince. That evening he brought Mr. Renauldi, purported to be Victoria's best, who witched another spot.

"At sixty feet," the rotund Italian predicted, "there es' beeg flow." He was so serious and sincere that our spirits skyrocketed.

While Lester and Mr. Renauldi amused themselves dropping sticks of dynamite down the dry hole, hoping to open a seam of water, I hacked a road through the bush to the new site. The explosions shook the island and sent showers of coal hundreds of feet in the air—but to no avail.

Once again, the thump . . . thump . . . thump awakened us at daybreak. But now I could no longer stand idly by. From watching Bob, I thought I knew how to handle the rig. At noon, we started working in shifts and kept the drill going sixteen hours a day. Unfortunately, the equipment was old and worn. At forty feet, the cable broke and we spent hours "fishing" for the bit. Blue shale dulled the cutting edge. It had to be unscrewed, heated, and pounded back into shape. Progress now came in sweat over the forge, and in inches instead of feet.

Then the blow fell. The bailer came up dry at sixty feet. "He could have been off a little," Bob said to soften the blow. "Might as well keep going. She could come in any time."

In spite of his optimism, our spirits plummeted. Nothing could crowd out the dismal fact that Wallace was waterless.

The next day confirmed the fact. Jeanne and I gazed gloomily at the rig, then at each other. We were despondent. Time was running short. The Gamons would arrive on Monday; it was too late to notify them not to come.

In silence we drifted back to the house. The damnable irony of it all, I thought. Our dream a failure—all on account of a little water. But now, as I look back, I know I had only myself to blame. A sea-lander must live entirely by his wits. I had failed to use mine.

I flung myself into a chair, boiling with anger and bitterness.

"Looks like we've had it." Jeanne's eyes were full of tears. "Damn it!" I cried, springing up and pacing the floor.

"Over two hundred acres and no water? I can't believe it.

There must be."

"I know it's a bitter pill, Dave, but we've got to face it: Wallace is bone-dry."

I didn't answer, for I was thinking of the dozens of fruit trees Chivers had planted. He must have had water. Plenty of it, to start them. Then the old man's words came to me. "Keep lookin', laddie. Ain't nuthin' the islan' don't have."

I turned to Jeanne. "Bob goes off for the weekend tomorrow, doesn't he? All right, we'll arrange to rent the rig and work it ourselves. This divining is a bunch of hokum." I picked up the geology book that had arrived in the mail the day before. "According to this, the most likely spot for water is in the meadow. We'll try it there."

The next day we stopped the drilling truck in the middle of the meadow. The dry grass rippled like wheat in the warm wind. Around us, firs stood so tall and straight that they strained the eyes. It was like a golden lake sunk in the forest.

"It's a natural basin," Jeanne remarked. "There's bound to be water here."

Excitement gripped us as we raised the tower and lowered the drilling bit. Jeanne bent down and kissed the ground. "Just for luck," she explained with a grin. The motor roared and the bit drove into the parched earth, the cable clanging against the pulleys. My hand steadied the cable as the huge drill bit rose and plunged, slamming the ground thirty tinges to the minute.

"I'll take the first shift," I said. "You better stick around the house in case a boat comes in." We had agreed to take turns and work around the clock, but I knew I would never stop as long as I could stand.

At dusk, Jeanne brought sandwiches, coffee, and a dish of wild strawberries. We sat on the ground and ate, almost too tense to speak. The strain and lack of sleep showed on our faces.

"How deep are we?" she asked.

"Twenty-four feet. I'm sure glad it's sandstone—shale is the devil on bits."

"You look tired," she said. "Let me take over for a while." I was, but I shook my head. "This is no job for a woman. Maybe later."

The sun dropped like a red ball behind the firs. Thump . . . thump . . . thump . . . by lantern light. Diamonds twinkled in the sky. A cool breeze swept across the grass with the smell of fir needles. I pulled the sleeping bag around Jeanne's shoulders. What a gal! She preferred keeping me company on the cold, shaking ground, to a warm bed.

I returned wearily to the box next to the drill. At 3:00 A.M., the cable marker showed thirty feet. Kiki sat beside me, and I fondled her head. My eyes were heavy, my feet numb with cold. The rhythmic thumping made me drowsy. I wished the motor would stop so that I could fill it with gas. And I began to think about how strange it was to be sitting on an island in the middle of the night, probing the earth for water.

Perhaps I dozed off. Suddenly it was broad daylight, the cable was rattling, and there was a slushing sound in the hole.

"Jeanne!" I cried, shaking her awake. "We've struck it!"

Excitedly we pulled up the drill and lowered the bailer. The pulleys groaned. The bailer came up spurting precious gray water. We whooped and hollered.

"Wait!" Jeanne said, very serious. "It might be salty. We'd better taste it."

We did—and threw our arms around each other. Then, dumping the bailer on the ground, we Indian-danced in the mud.

CHAPTER 18

Beef Hearts for Bruno

WHEN THE GAMON STATION WAGON, LOADED FOR A SIX-MONTH safari, rolled up to the wharf, I had great misgivings. What would they think of antiquated Bertha? With two small girls and so much gear, they might be afraid to trust her. Wharf- and cargo-scarred, she was anything but a plush resort launch. She didn't even have a cushion.

Mr. Gamon looked like a banker in his spotless black suit—lean, soldierly erect, with shrewd, intense eyes. A man one would never call by his first name. I had visions of his saying, "Good lord, you don't expect us to get in that tub!"

Instead, on the float, he said in a not-unfriendly voice, "Some boat you've got there, Mr. Conover." He handed me grips and duffel bags. "What will she do—four or five knots?"

The way the sea was kicking up, I doubted she'd do that. I was too flustered, loading boxes, suitcases, and toys and wondering where everybody would sit, to notice his familiarity with boats.

Halfway across, the northerly chop sprang into a squall. Leaping and pounding, Bertha shattered the wave tops. Green water slopped over the gunnels. My eyes nervously scanned our guests. Mrs. Gamon, a bright, attractive southern belle, wore a confident smile; Sally and Christine sat giggling and ducking the spray. Perched on the luggage, Mr. Gamon faced the seas, puffing his pipe with sober equanimity.

Then it happened. A giant wave broke over the bow and swamped the motor. I was dismayed. Clothes and luggage were soaked and the boat rolled crazily. I crawled to the motor, fearing our guests would depart in rage. Oddly enough, they calmly helped lash down the luggage and dry out the ignition.

"Don't worry, Mr. Conover," one of the youngsters said. "This has happened before."

And no wonder. They were a Navy family, used to puttering around in small boats; the man, Captain John Gamon, Commander of a Naval Air Station. It was a wonderful case of beginner's luck.

Jeanne and I were awkward hosts. Not having been around people for a year, we were horribly self-conscious—worried about our worn clothing, our shaggy haircuts, and whether we would say and do the right thing. We had no entertainment for guests. No radio or TV—just a badminton net strung between apple trees and two pits dug in the grass for horseshoes. Would people have enough to do? We had no insurance, either. What if someone broke a leg—or drowned?

All summer, we were a bundle of fears. We were afraid guests would depart, bored and disgruntled; or that someone would steal their cars. We worried about the fire hazards of cigarettes and gas lamps, and about how to get food on rough days and how to keep it without refrigeration.

Fortunately, the Gamons were ideal guests—kind, sympathetic, and appreciative. It was a good thing. Upon them fell our worst boners. The first evening—choking and coughing—they fled into the yard: we had forgotten to remove the chimney cap. Our faces were even redder the next morning when they limped toward us, bleary-eyed and clutching their backs. We had also forgotten about the sheet of plywood under the mattress.

However, things turned out fine.

The third day, the Commander asked me to call him John.

The Gamons were a bit like us. They liked out-of-the-way places, didn't mind solitude or gas lamps, and were omnivorous readers. After a few days, they announced they wanted an island of their own. We explored one listed by the village agent. Then they thought it over.

Next day, the Commander drew me aside. "Helen and I have decided there's no point in buying an island as long as we can come to Wallace. You've spoiled us. There isn't an island anywhere as lovely as this."

That night we went to sleep with a smug smile.

The last day, when the Commander brought out his checkbook, he said, "You kids have something here. A home, the island, an outdoor life, each other. What more can you ask" He held out a check. "This includes a deposit for next year."

A few weeks later, we were not so lucky with the Blodgetts.

Their letter had stated they'd arrive by air. Tense and excited, we waited on the float as the small seaplane taxied up. Avoiding our gaze, an elderly woman sat stiffly at the window with a dour expression.

"This doesn't look good," I whispered to Jeanne.

In a pall of exotic perfume, Mrs. Blodgett squeezed from the cabin and eyed the greenery with a suspicious glare. She was straight from the pages of the Boston social register—a Prussianly erect dowager with tiny fluttery eyes, breasts as large as basketballs, and a double string of pearls dangling over them like ice from a precipice.

She extended her hand with serene matronliness. "Why, you're Mr. and Mrs. Conover, aren't you? I thought you'd be much older."

We nodded dumbly. Not knowing what to say, we beamed welcoming smiles.

"So dear of you to meet us," she went on politely, nervously fingering her pearls. "You know, this was Harold's idea. He always wanted to get off into the wilds. Where everything is primitive, you know. I thought a small island might be best—one, you know, where he can't get lost."

Harold, a small, rather frayed-looking businessman, smiled at us faintly. He was clutching her white gloves and a makeup case in one hand, a gray toy poodle in the other. The nerve, I thought—a dog, of all things!

"We're traveling light," Mrs. Blodgett continued. "You know air travel. You can take only the bare essentials."

The pilot set a bag of groceries and two small cases on the float. They were traveling light indeed.

As the poodle dashed up the gangway, another plane entered the cove.

"Oh, yes—I forgot to tell you, Mr. Conover." Mrs. Blodgett pointed. "That's our luggage coming now."

Jeanne and I exchanged glances. This was a job for the pushcart!

"Beautiful... absolutely beautiful!" Harold seemed to appreciate our cove with childlike awe. "A whole island to ourselves. How marvelous! This is real living."

It was almost noon. We had been up since four—scrubbing floors, dishes, windows, and walls.

"But what do you do all winter?" Harold went on. He was one of those nervous, hurried, terribly keen Americans who always have to be doing something.

I was about to answer when Mrs. Blodgett thrust a parcel at me. "Please put these in the freezer, Mr. Conover. They're beef hearts for Bruno—the only thing he'll touch. He's probably starved, the poor dear! He hasn't had a thing since breakfast."

I didn't have time to explain.

Kiki pounced from the salal and dug her claws into Bruno's back. He fled yelping up the path. Dropping everything, we dashed after the screaming animal.

"Stop that cat!" shrieked Mrs. Blodgett, waving a shoe. "Bruno will be murdered!" I couldn't think of a better idea.

When Bruno was finally safe in her arms, Mrs. Blodgett faced us indignantly. "Mr. Conover, please keep that cat inside after this."

The shack had never looked lovelier. Window boxes bloomed with petunias, and dahlias hugged the walls in hedges of orange and gold.

As we began to unload the bulging pushcart, I heard a gasp. "Oh, dear!" cried Mrs. Blodgett. "There must be some mistake."

"Nope, this is it," I replied stoically, putting the luggage inside. "You'll find everything you need. Linens are in the dresser, the woodbox is full, and we'll show you later how to light the lamp. Call on us anytime." And I pointed, "We're in the cottage over there."

Our simple life suddenly turned into bedlam. Mrs. Blodgett complained the wood stove gave her asthma, the water gave her "the trots," and she swore repeatedly that our yellow-legged mosquitoes carried malaria. Food, wood, and laundry became insurmountable problems. She burned herself with the flatirons. Our medicine cabinet emptied at her feet.

"I'll be scarred for life!" she bellowed. "You've no business leaving those dangerous things around."

Harold got splinters from packing wood, lumbago from digging clams. They had to have powdered bread crumbs to fry their fish, and beef hearts for Bruno precisely at four each afternoon, which, regardless of the sea, required a daily trip to the freezing compartment of Myrtle's refrigerator.

When I casually mentioned that they were drinking raw milk, Mrs. Blodgett flew into a rage.

"*Raw milk!* Oh, my god, Mr. Conover—don't you know cows carry disease?"

To Mrs. Blodgett, fresh air was as painful as silence. Windows and doors were clamped shut and holes in the cooler were plugged. Every night we were commandeered for bridge. The room was close and hot, her exotic perfume overpowering. Neither of us played anything but canasta. The result was Dunkirk every hand. We staggered to bed at daybreak, reeking of perfume and cursing the resort business, wondering what was wrong with our heads.

The first few days Kiki was like a caged beast, snarling and scratching the door when she heard the poodle outside. Bruno left no part of the island unscathed. His unbelievably small bladder had an infinite capacity to produce at every stop. The dahlias wilted, and stains appeared on the white horseshoe posts. The wild juniper that we had transplanted and nursed through the drought suddenly collapsed and died. None of this bothered the Blodgetts, but I told Jeanne, "I know one thing that *will.* Their *bill!*"

Harold was never out of our hair.

One morning he stalked from the house like an English country squire in tweed knickers, beige corduroy jacket, and a bright red cap the shade of his squatty, sunburned nose. Across his shoulder was a double-barrel shotgun, and a weird-looking horn hung from his neck. He disappeared briskly among the trees.

A moment later, the quiet was shattered by the long wailing cry of the huntsman's horn. There was a flapping of wings and a restless agitation of fir branches. Boom! The crows rose in a black cloud.

Rushing outdoors, I made my way cautiously into the woods. Harold emerged from the thicket, smiling smugly and dangling a handful of crows.

"Look! All in one shot." He grinned, shoving them under my face.

I brushed his arm aside. "I'm sorry, Mr. Blodgett"—I tried to control my temper—"we don't permit shooting on the island."

"Not even crows?" His voice was shrill in disbelief.

"No, not even crows," I repeated sternly. "This is their home as well as ours. We never bother them."

Harold looked baffled, and his face reddened. "They're monsters! I can't stand them. You've got thousands—why don't you let me get rid of 'em?"

I repeated my warning more emphatically. "There'll be absolutely no shooting on this island. None whatsoever."

It didn't stop Harold. That afternoon he disappeared in the rowboat.

While we were eating dinner, Mrs. Blodgett rapped on the door. "You-hoo! Have you Conovers seen Harold?" We shook our heads.

By seven o'clock Mrs. Blodgett was frantic. "Won't you go out and look for him?" She tugged at her pearls in distress. "He's probably lost, the poor dear."

I felt like saying "Good," but I just sighed. "All right. If he isn't back in an hour, I'll try to find him."

An hour later, it was almost dark. Still no Harold.

"Have you any idea where he went?" I asked Mrs. Blodgett.

She thought, biting her lip and clutching her pearls. "He said something about bagging a seal."

"Seal rocks!" Jeanne turned to me. "That's where he is. Those reefs off Princess Harbor."

Sure enough—he was standing on the rocks, soaked and shivering, waving his arms. But no sign of the boat. Or the gun.

"I've been yelling for hours," he croaked as we came alongside. "I did the damndest thing... can't understand it." Then he explained what had happened.

He had seen two seals sunning themselves on the rocks and quickly raised his gun and fired, but in his excitement he forgot the wide range of his shot, and the hull was peppered with holes. He rowed frantically toward shore, the sea bubbling up around him, and swam the last fifty yards with the bowline wrapped about his arm.

Harold shook his head. "Can't understand it," he moaned while we towed the boat home. "How could I do such a stupid thing?"

I wondered too—how much to tack on his bill.

The fifth day, outfitted like a beekeeper out to collect honey, Mrs. Blodgett ventured outdoors. Her face and legs were swathed with mosquito netting, and a pail hung from her arm. Grim determination gleamed in her eyes. We guided her to an enormous patch of blackberries, then retreated to the cottage for lunch. An hour later we were startled by a loud shriek.

"Good!" Jeanne exclaimed. "A yellow jacket has got her."

She came screeching down the path, the netting streaming behind her like a sail and the pail spewing berries along the way.

"A rat!" she cried. "It ran out of the bushes."

"That's a mink," I chuckled, trying to keep a straight face. "You probably scared it to death."

With a gasp of relief, she suddenly felt her chest. "My pearls! They're gone!"

We spent the rest of the day on hands and knees, searching for them, Jeanne muttering, "Who suggested blackberries?" That evening at cards, we were informed Mrs. Blodgett had not worn her pearls.

The Blodgetts' departure came sooner than we dared hope. There was a knock on the door at noon, and I got up from the table with a feeling of apprehension. "What's up now?"

"My plumbing is out of order." Harold's pinched face stared at me. "Would you go over and call a plane?"

I was confused. What had our plumbing to do with a plane? Had I made some insufferable boner? Did Mrs. Blodgett fall into the septic tank?

"Isn't there something I can do?"

"Not this time," Harold replied urgently. "My bladder has gone haywire. I've got to get home to my doctor."

The channel was alive with whitecaps, but somehow I didn't mind risking my neck for a plane.

Late in July we had a honeymoon couple, the Martins—they'd been married that very noon. We took them on a moonlight ride—had broiled oysters on the beach. The way they held hands made us feel ancient.

The next day when they failed to put in an appearance, we were not surprised. On the third day, Jeanne came down to the wharf where I was working.

"They haven't come out yet," she said. "Suppose something's wrong?"

I grinned. "Maybe they're too weak to get to the door."

As the summer progressed, the floats began to pay off. The reef markers and signs on the points helped. Three American yachts tied up for the night. They were more like floating cocktail parties. Along toward midnight we couldn't stand it any longer.

"You better go down and tell them not to make so much racket," Jeanne said.

Before long, we began to wonder whether yachtsmen were a blessing or a blight. But we needed the money. In June, we had barely broken even.

People from the boats roamed everywhere, peeking in windows, snooping in the shed, playing horseshoes and badminton (we finally hid the racquets), and cluttering the trails with candy wrappers. We were constantly interrupted; repeatedly asked: Do you have ice? Where is the dining room? The post office? Which way are the restrooms? Can you spare some gas? It grew increasingly difficult to take time to chat, even to be courteous. After a long ordeal with an elderly couple who wanted to tell us their life history, I grumbled to Jeanne, "Thank goodness, they're only here for the night."

Small boats were often a problem. When outboard cruisers docked, kids poured out. The island reeled with shrieks and shouts, the yard became a melee of bare arms and feet. Kiki took refuge under the house. Horseshoes landed in the bush, and gravel from the path peppered the lawn. If we were lucky, we survived with only a broken window.

"For a dollar," Jeanne declared, "it must be a bargain to get the little monsters ashore."

Nevertheless these sea-jockeys were an amusing breed. To them, a chart was a map, the sea just a street. Instead of going astern, they backed up. For starboard or port, it was left or right. They docked as if they were pulling into a gas station at Wilshire and Fairfax; and tanking up, we discovered, was what they did most of the time. We chuckled. Their vacations were merely itineraries between gas docks.

As more boats came in, we realized that yachtsmen fell easily into types. Those with puffed glassy eyes we distinguished as "bottom-happy boys." They navigated by depth-finder instead of by chart. We felt sorry for them. Among these fir-lined waterways, if one is overly conscious of the bottom, he rarely enjoys the top.

We felt even sorrier for the "lonely soul." As soon as he docked, he flicked on the radiophone full-blast: "Hello, Sam. This is Lollipop calling. Can you read me, Sam? OVER. Hello, Dingle. This Mary Anne. Where are you, Sam? OVER. Can't read you, Dingle. Come in again. OVER. Hello, Dingle. I read you swell. Can you read me now?" And so it went on into the night.

For the "odd-ball," we had a mixed feeling of pity and fear. The first one zoomed into the cove atop his flying bridge—bare-chested, as gorgeously bronzed as a Greek god, in white trunks and braided cap—and shouted: "Which way is it to Alaska?" We never knew whether the guy was incredibly brave or stupid—or incredibly drunk!

And, lastly, we shall never forget the "lunker" whose love for the sea was the only thing that kept his tub afloat. When he wasn't mending its creaky timbers, he was down on his knees bailing it out. He knew nothing about valves, points, and fuel pumps. Consequently his holiday was largely spent tied to our wharf. A natural pest, he never had enough tools. It started with a wrench, friction tape (just a piece); then a board or two—a cup of flour—a cube of butter. But, oddly enough, we hated to see him go.

Jeanne and I were always on the go, knocking ourselves out trying to please our guests. Sleep was a winter luxury; so were three meals a day. For trouble constantly knocked at the door. We never knew whether to grab the plumber's snake, the fire extinguisher, or the crowbar. People considered a toilet an automatic refuse disposal, fire was a constant threat, and kids were always getting themselves locked in the bathroom.

Before the season was half over, we were bleary-eyed and exhausted. We couldn't understand it. How did other resort owners manage? A kind guest tipped us off one day. "You give too much of yourselves," she said. "You can't please everyone." Still, we tried. We resolved never to be caught idle—or wasting time on such ordinary things as eating, sleeping, washing, or going to the bathroom. Nothing is worse for a guest's morale than seeing his host without anything to do.

Between being polite to yachtsmen and keeping guests happy, entertained, and supplied with grub, we laid the pipeline, installed the pump, painted outdoor furniture, mowed grass, bailed boats, chopped wood, weeded the garden, limed the badminton court, and kept the trails open with the machete. In spare moments, we emptied garbage, hunted horseshoes, raked leaves, oiled floors, opened oysters for guests, gave barbecue parties, and answered questions: Don't you ever get lonely? Bored? Or frightened? How did you get here? Aren't you dreadfully out of touch? How do you stand the winter? The way we felt, winter couldn't come quick enough!

There was always the unexpected—the shipwreck of *Wayfarer II,* a forty-foot motor sailer, on Panther Point; the day we took off to help the customs inspector search for a missing freighter; and the time Mr. Gerritt knocked on our door at midnight and gulped, "My wife—I think—she's going to have a baby!"

It was if a dentist had struck a raw nerve. "Not here, Mr. Gerritt!" I panicked. "I'll get the boat." Fortunately, it was a false alarm, brought on by eating too many oysters and clams.

As summer rolled by, the way guests enjoyed themselves exceeded our fondest dreams. They came to us tired, nervous, and frazzled. In a week (the minimum for an American to unwind), our big green ship worked miracles. The freeway lines vanished from their faces and the pressure of the dollar from their eyes. The silence of the sea, sky, and trees was therapeutic. It made us feel good to see them leave more hopeful, healthier, and happier.

It seemed to us people were more real, more human, behind a fishing rod, in a boat, or hiking in the woods. At a cocktail party, a business lunch, or behind a desk, we meet only their masks. But Nature dissolves pretense, insincerity. We discovered early it takes only a few trees to whittle a man down to size.

It was wonderful to see our dream taking shape. The few moments we had together, we gushed with future plans: more cottages, a store and recreation room, gas and water at the dock, a sea-diked swimming pool, benches and picnic sites around the island. "Quality instead of quantity," Jeanne insisted, "so Wallace won't lose its charm and become just a public park."

However, our optimism was short-lived. It grew daily more obvious that we had badly misjudged our overhead. We desperately tried everything to make money. We wrote yacht clubs and newspapers to attract more boats. Sold salal and baby firs in oil cans, picturesque driftwood to collectors, and oysters by the pint. Guests left with boxes of plums and apples. We even sold sand dollars, moonshells, and dried sea urchins. The oyster-shell ashtrays went well, too.

A week after Labor Day, we counted our take. Our financial report was anything but impressive:

Cottage	$ 825.00
Moorage, boat rentals	240.50
Miscellaneous	72.25
	$1,137.75
Expenses	844.25
Net profit	$ 293.50

We stared gloomily at each other.

"That won't even cover the interest," I groaned. "Maybe we'd better withdraw in dignity before we break our hearts."

CHAPTER 19

The Great Gamble

"WHAT KIND OF TALK IS THAT?" JEANNE CRIED! "We still have twenty-three days. You'd better head for town and not come back until you get the money. I'll be all right—I've got those pears to do. After all, we want a full larder this winter."

Her words shot fresh hope into my veins. "Yes, ma'am," I answered, smiling. "How would you like the thousand dollars? In tens, twenties, or fifties?"

At daybreak I took off for Victoria. I went to more banks, more loan companies. Alas—the answers were always the same. Canada was still gripped by postwar austerity.

Even Lester shook his head. "Honest, Dave, I don't know what to suggest. Every cent I've got is sunk in this business. Believe me, I wish I could help." He butted his cigar and, after a moment, went on seriously, "There's one thing you can do."

"What?"

"Log the island."

"Turn Wallace into a rockpile?" I exclaimed, amazed at his proposal. "No—we're too fond of our trees."

My heart was heavy as I set out for home. I was certain that only a miracle could save us now.

For several days we went through the motions of living. Night after night we lay awake in the moonlight, staring at the cove, wondering what we could do. On Friday, we were working in the garden when a stranger suddenly appeared carrying a sack over his shoulder. He stopped short in surprise.

"Gosh, I didn't know any one was here," he apologized. He was gathering cascara bark, he explained, and had thought that the island was uninhabited. We invited him in for coffee.

His friendly eyes surveyed us and our simple cabin. "Suppose you folks know you've got some fine timber. Worth plenty, too, them big firs. They'll just rot and die if you don't take 'em out. It gives the young stuff a chance."

We hadn't been aware of this—only of their magnificent beauty. To log the island, we believed, would destroy it.

"Dave, this might be the answer," Jeanne said.

With hearts pounding, we questioned him further. How was timber sold? The market value. The ways the island could be selectively logged. Yes—it could be done! But time was running short.

We hurriedly sent away for books on logging. Then I remembered the forestry book we had, dug it out, and studied the chapter on "Timber Cruising." Following its directions, we began cruising the island with tape, log scale, and paint bucket, marking the trees we were willing to sell, taking care not to spoil the island's beauty. When we totaled our figures, there was more than enough to meet the $1,420 payment. We were jubilant.

But not for long. An hour later, as she was going over the fine print of the mortgage, Jeanne let out a groan. "Listen to this: NO TREES ARE TO BE CUT OR REMOVED WHILE THE CONTRACT IS IN FORCE."

"You mean, we've got to dig up over fourteen thousand dollars before we can sell one tree?" At Jeanne's nod, I sank into a chair. "We're right back where we started."

We checked the timber again. There was not nearly enough to clear the whole mortgage.

This time, I was convinced there was no way out—we had used up our share of miracles. Nine days were left. I knew it was wrong and selfish to make Jeanne suffer needlessly. My mind said Quit but my heart said Keep on. Still, I was not going to beg for more time. What good would it do? We needed capital—money to build more cottages, money to live on, money to pay off our debts.

"Guess I better dig out that pin-stripe suit," I said finally. "You know where it is?"

Jeanne laid her hand on mine. "Doesn't matter. I prefer the way you look now."

"What do you mean?"

"I wasn't figuring on leaving—I didn't think you'd quit so easily."

I looked at her in surprise. "We're licked, Jeanne. We haven't a straw to hang onto."

"Maybe you are, but I'm not. What's got into you, anyway? We've still got nine days, haven't we?"

"You really... really want to hang on?"

She nodded.

"I don't want to leave either. But you—you've been through so much."

She threw her arms around me. "Darling, that pin-stripe suit wouldn't fit you."

Later, we were sitting at the edge of the cove when Jeanne came up with what I thought was a foolish idea: "Why don't we round up more timber?"

"What with, silly girl?"

"How about an option?"

Her simple suggestion changed a disaster into a plan. The Secretary Islands flashed into my mind. We cruised them, and our hopes soared. There was more timber than on Wallace, roughly a million feet. On pins and needles, we phoned the owner.

Yes, he was willing to sell. How much?

My heart sank.

He repeated it: $25,000, hard cash. Wow!

With our fingers crossed, we plunked down fifty dollars for a two-week option. Then, after a hurried reappraisal of the islands, we suddenly realized that an option was not enough.

"I don't like it," Jeanne said. "We need more protection. What if the timber buyers stall? They could cook up a deal with the owners, to go into effect when our option expires."

It meant only one thing—we had to sign an agreement to purchase. This kind of financing frightened me. It sounded too complicated, too risky. If it didn't work, we'd...

"You mean that we should buy more islands?" I protested to Jeanne.

"Go into more debt? So we can pay off the debts we have already?"

"Why not? It's just as easy to lose three islands as one."

I was uneasy. We were stepping too far out on the limb. If this came off, it *would* be a miracle.

On Thursday, we signed the agreement to purchase.

We now owed $39,420. In the bank, we had $187.10.

The same afternoon, we rushed "Timber for Sale" ads into the Vancouver papers, hoping to attract cash buyers from the largest lumbermills. That evening an uncomfortable fact began to badger me. Island timber was mostly # 3 grade, much of it windblown and conky.

I turned to Jeanne. "What if they don't go for it? Don't offer enough? We'd be stuck!"

"That's a chance we've got to take," she said gravely.

But it bugged me. I wasn't one for leaving things solely to chance. A plan began to flicker in my mind, and I got out my old college text and spent the night reading *Psychology and Life.*

In the morning, we were flushed with excitement. Timber buyers began to arrive by plane—tough, shrewd, forest-bred men in laced boots and leather jackets, their lunches tucked under their arms. I greeted each one at the wharf. Then, by boat, I scattered them separately on the three islands so that they would meet unexpectedly in the bush, hoping they'd think, "Well, if Charlie is here, this must be a really big show!" It was timberman against timberman.

Bids came in thick and fast.

On September 29, a day before the deadline, we accepted the highest cash offer, enough to pay off Wallace completely, satisfy our creditors, and give us two beautiful islands as well.

In October, we burned the mortgage, scattered the ashes at sea, and offered our humble thanks to the Lord. For weeks we were a little dazed, unable to believe that our kingdom was truly ours.

With Wallace free and clear and the resort a success, we thought we had used up our share of good fortune. Two years later, when the Secretaries were sold and more cottages were under way, yet another prayer was answered: A little sea-lander arrived on the scene—a son, David, to make our lives complete.

Twenty years have passed since those first difficult greenhorn days. Each one has brought its problems and pains, but also special joys deep and rich enough to make them pale in memory. Jeanne and I have found peace here. The sea and the resort provide all the tension we need—often a bit more. During the tourist invasion, we bank enough of society to last the winter; in winter, we bank enough privacy to carry us through the busy summer. We are not wealthy in worldly goods. We still heat our simple home with wood, and we have little entertainment we don't make ourselves. If we have few amenities, we have a full measure of beauty, companionship, and freedom.

Freedom, I believe, is fighting a rearguard action these days. Where it's most talked about, men are least free. And where there are fewer men, there is generally more freedom. A man is his true self only when he is alone. "If he does not like solitude," a great German once said, "he will not like freedom. For it is only when he is alone that he is truly free."

To me, a necktie is a chain; an old pair of pants, freedom. If this sounds foolish and naive, at least I am free to be myself—and free to enjoy to the fullest my uniqueness as a man.

When you shoulder the day and share its wonders with those you love—whether you are scrubbing floors, pulling in a fish, or weeding the garden—there's much to be grateful for. And, as Jeanne says, "Nothing holds a candle to growing your own food, building a root cellar, or hand-feeding a lamb." An outdoor life and a proud pleasure in one's work—what greater joys are there than these?

Today, civilization presses hard against these fir-clad shores. Wallace is no longer a resort. The cabins and grounds are now the property of individual owners, some of whom were former guests. It is, for them, a measure of a dream—and they find it as much a paradise as do Jeanne and I.

We know an island is not everyone's cup of tea. But we like to think our story proves that a fantastic dream can come true—if your dream is strong enough. The tragedy is that so many of us get sidetracked or lack courage. So get busy with all your heart. Let no one tell you there isn't magic in a dream.

Publisher's Note

We want to thank David "Davey" Conover, Jr. for his assistance in making this edition of *Once Upon an Island* possible. He provided the high quality original photographs, taken by his father, from the first printing of the book in 1967. He also provided us with 1967 newspaper reviews of the book.

David Conover, Sr. passed away in 1983 and Jeanne passed away in May 2003. David Jr. is currently the proprietor of Father's Country Inn Bed and Breakfast located in Kamloops, British Columbia, Canada. You can find out more about David and his family from his web site: www.dconover.com.

Also available from San Juan Publishing:

The Light on the Island by Helene Glidden

San Juan Publishing
P.O. Box 923
Woodinville, WA 98072
425-485-2813
sanjuanbooks@yahoo.com